Prostaglandins, leukotrienes,
and the immune response

Prostaglandins, leukotrienes, and the immune response

JOHN L. NINNEMANN
University of California, San Diego

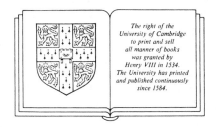

The right of the
University of Cambridge
to print and sell
all manner of books
was granted by
Henry VIII in 1534.
The University has printed
and published continuously
since 1584.

CAMBRIDGE UNIVERSITY PRESS
Cambridge
New York   New Rochelle   Melbourne   Sydney

Published by the Press Syndicate of the University of Cambridge
The Pitt Building, Trumpington Street, Cambridge CB2 1RP
32 East 57th Street, New York, NY 10022, USA
10 Stamford Road, Oakleigh, Melbourne 3166, Australia

© Cambridge University Press 1988

First published 1988

Printed in the United States of America

Library of Congress Cataloging-in-Publication Data
Prostaglandins, leukotrienes, and the immune response / John L.
Ninnemann
p. cm.
Includes bibliographies and index.
ISBN 0-521-33483-7
1. Prostaglandins – Immunology. 2. Leukotrienes – Immunology.
3. Immune response – Regulation. 4. Immunologic diseases –
Pathophysiology. I. Title.
[DNLM: 1. Arachidonic Acids – metabolism. 2. Immunity, Cellular.
3. Leukrotrienes B – metabolism. 4. Prostaglandins – metabolism.
5. SRS-A – metabolism.   QW 568 N715p]
QP801.P68N55 1988   87–35429
612'.405 – dc19   CIP
DNLM/DLC

British Library Cataloguing in Publication Data
Ninnemann, John L.
Prostaglandins, leukotrienes, and the
immune response.
1. Animals. Immune reactions. Role of
metabolites
I. Title
591.2'95

ISBN 0 521 33483 7

To my mother and father
Bernice and Milton Ninnemann
whose sacrifice provided me with an education
and opportunities they never enjoyed and whose
love provided me with the strength and spiritual
values necessary to live in a difficult world.

CONTENTS

FOREWORD

Single-author books are becoming somewhat of a rarity in scientific literature. Far more common are the multiauthored loosely edited works focused on a single topic. Such publications have their uses, especially in giving the reader first-hand accounts of relevant research findings. On the other hand, multiauthored works frequently fail when it comes to providing a synthesis, an overall picture of a complex field, wherein seemingly unrelated findings are placed in the "big picture." This is an important goal for any scientific book, especially one devoted to a field as rife with seemingly contradictory data as is the area of eicosanoids and immune function. Dr. Ninnemann succeeds admirably in meeting this goal. He has managed to integrate into a readable text an enormous amount of data, results obtained *in vitro* and *in vivo* in animals and in humans.

How important are arachidonic acid metabolites in the normal immune response? We assume that cyclooxygenase metabolites are not critical for normal functioning, because both humans and animals seem to do quite well on long-term treatment with cyclooxygenase inhibitors. On the other hand, overproduction of cyclooxygenase metabolites and/or changes in sensitivity to these metabolites would appear to play a role in the disturbances in immunoregulation associated with several diseases. Less is known about the role of the lipoxygenase metabolites of arachidonic acid, but it is clear that one or more of these lipoxygenase metabolites must play an important role in normal immune function, because all the lipoxygenase inhibitors recently developed are powerful immunosuppressants.

Dr. Ninnemann has divided his text first by specific target cells, devoting a chapter to the effects of eicosanoids on lymphocyte function, on monocyte function, and so forth. He then turns to a more global discussion of the participation of eicosanoids in various diseases. Throughout the entire book the reader benefits from Dr. Ninnemann's encylopedic knowledge of the field and from his continued attempts to provide an overall integration of the experimental data reviewed.

James S. Goodwin

Medical College of Wisconsin
Milwaukee, Wisconsin

PREFACE

In his revolutionary book *The Structure of Scientific Revolutions,* Thomas Kuhn describes the nature of scientific progress, and how new, better answers to scientific problems become validated and accepted. Progress, he suggests, comes from the construction, and frequent revision, of models or paradigms that explain sets of information. These paradigms are the underlying assumptions of any scientific field. It is significant that Dr. Kuhn's philosophical constructions continue to attract attention and have helped organize thinking in a number of complex fields, including immunology.

It has become clear to me in the preparation of this monograph that a new immunological paradigm is emerging that relates to the participation of the products of arachidonic acid metabolism in the immune response. It is now clear that the prostaglandins (particularly PGE_2), the leukotrienes (particularly LTB_4), and probably the lipoxins act as primary and/or secondary mediators in a wide range of immune functions and disease states. Synthesis of this paradigm began with the observations of many individuals, then progressed to reviews on mediator involvement in specific diseases or cell types, and finally to recognition of the ubiquity of the prostaglandins and leukotrienes in normal and abnormal immune responses.

Our understanding of the exact nature and importance of the participation of these mediators in immune function is in its infancy, however, and some important basic questions have not yet been answered to everyone's satisfaction. Before any system can be manipulated to clinical advantage, it must be understood. It is hoped that the information assembled in the following pages will contribute toward that understanding.

I am greatly indebted to many friends and colleagues who helped me in the preparation of this monograph. A few of these individuals deserve special mention and thanks, including Jim Goodwin, for his patient review of each chapter as it was written and his many suggestions and wise counsel; Murray Mitchell, for his help with the manuscript and for making his library available; Doug Green and Jane Shelby, for their review of the final draft; Otto Plescia, for very helpful discussions during his visit to San Diego; and Marie Foegh, Guy Zimmerman, and Ivan Bonta, for sending large volumes of important material for inclusion in appropriate sections of the book. Because of the skill of Audrey Threlkeld with Wordstar and her good humor, I and the manuscript survived the many rewritings necessary to satisfy my reviewers and editors. And with the help of Lisa Dressel, many elusive references were located, and letters written seeking permission to reprint the many excellent figures included here. Finally, because of the never failing patience, encouragement, and inspiration provided by my wife and best friend, Laura, this project was completed with enthusiasm, and, miraculously, on time.

1

A brief history and introduction

The products of arachidonic acid metabolism, including the prostaglandins and the leukotrienes, are surprisingly versatile compounds, which participate in an extraordinary variety of normal physiological processes, such as maintaining blood pressure and body temperature, protecting organs from damage caused by disease, traumatic injury and stress, and regulating parturition. In addition, an imbalance in these same metabolites has been implicated in shock and a wide variety of disease states including arthritis, malignancy, and allergic disorders.

In spite of their great importance, most of the lay world and a large portion of the scientific and medical communities, were not introduced to the prostaglandins until the announcement of the Nobel Prize in Medicine for 1982. Swedish chemist Sune Bergström, his colleague Bengt Samuelsson, both of the Karolinska Institute, and British pharmacologist John Vane of Wellcome Research Laboratories received Nobel recognition for their part in determining the structure and the biological role of the prostaglandins. Bergström's pioneering contribution was the discovery that prostaglandins are synthesized *in vivo* from dietary polyunsaturated fatty acids. Along with von Euler, Bergström was able to purify and further characterize the prostaglandins, which had been discovered 20 years earlier. The metabolic fate and disposition of the prostaglandins in the body were determined mainly by Samuelsson and his group. Vane demonstrated the ability of antiinflammatory substances such as aspirin to inhibit prostaglandin synthesis. Bergström, Samuelsson, and Vane also participated in the elucidation of many of the fatty acid derivatives, and the recognition that they often work as antagonistic pairs. One metabolite, for example, lowers blood pressure and another raises it. One dilates bronchi and a second constricts them. One promotes the inflammatory process, and another inhibits it. In 1973, Samuelsson showed that platelets produce thromboxane, which then participates in the

1

coagulation cascade. Vane identified prostacyclin in 1976, a powerful inhibitor of clotting.

The elucidation of these compounds, however, really began in 1930 with the work of Raphael Kurzrok and Charles C. Lieb, who studied the ability of human semen either to relax or to contract isolated strips of uterine tissue, depending upon whether the woman was sterile or fertile (1). Several years later, Maurice Goldblatt in England and Ulf von Euler in Sweden independently observed a similar phenomenon using human seminal plasma and extracts from sheep seminal vesical glands (2). Believing that the substance he had isolated was a product of the prostate gland, von Euler coined the misnomer *prostaglandin*. More recent work has shown that prostaglandins are not unique to any specific tissue, but instead are synthesized by a wide variety of cell and tissue types.

In the early 1950s, Vogt reported his observations of a highly active, apparently endogenously synthesized biochemical substance in the intestines, which he called *Darmstoff* (3). Several years later Ambache described a highly active lipid material in his experiments with rabbit iris, to which he gave the name *irin* (4). Welsh physiologist V.R. Pickles later described what he thought to be a singular compound, then called *menstrual stimulant*, which displayed the ability to activate powerful contractions of the uterus during menstruation; in 1965, a team of researchers at St. Louis University discovered a substance in the medulla of rabbit kidney, which lowered blood pressure (reviewed in 5, 6). All of these apparently unrelated biological events have since been attributed to the presence and action of the prostaglandins.

Leukotrienes were discovered during the elucidation of a mixture of compounds called the slow-reacting substance of anaphylaxis or SRS-A. SRS-A was first described in 1940 by Kellawey and Trethewie (7) to be a product of immediate-type hypersensitivity. When recovered from perfused guinea pig lung, this isolate was found to constrict smooth muscle tissue more slowly than histamine and, therefore, was named SRS-A. It is possible that Harkavy recognized the presence of a similar substance in the sputum of asthmatic humans as early as 1930 (8). SRS-A was later characterized as a unique "polar lipid" by Brocklehurst (9) and by Strandberg and Uvnas (10), distinct from the family of lipids identified as prostaglandins. Research, however, was greatly hampered by the extremely minute quantities of this material, as well as by substantial degradation of the compound during isolation attempts. The composition of SRS-A was eventually determined, however, by Samuelsson and colleagues, by Morris, Piper, and colleagues, and by Lewis, Austen, and colleagues to be a mixture of leukotriene C_4 (LTC_4) and its biologically active conversion products, LTD_4 and LTE_4 (11–13). Working from a knowledge of these substances, Radmark et al. definitely established that leukotrienes were synthesized from arachidonic acid, by way of the initial formation of hydroperoxyeicosatetraenoic acid (5-HPETE), then conversion to LTA_4, which is then further metabolized to either LTB_4 or LTC_4 (14).

With the elucidation of the prostaglandins and the leukotrienes as families of distinct, biologically active compounds, has come a great quantity of research activity to describe their natural occurrence and physiological roles. One of the more exciting and significant of these roles is the participation of the prostaglandins and the leukotrienes in the regulation of immune responsiveness. Unfortunately, our understanding to date is incomplete and conflicting data at times are hard to resolve. Because of this, author bias is very evident in the chapters which follow, in terms of the emphasis given to various bits of data, the hypotheses constructed, and the studies relegated to "conflicting data" status.

A BRIEF OVERVIEW OF THE IMMUNE RESPONSE

In the chapters which follow, the immune response will be described to consist of two basic components: the nonspecific and the specific immune response. While the specific immune response is capable of recognition and an amplified secondary response, nonspecific immunity refers to the protective roles of such varied systems as the unbroken skin and phagocytosis, which are not capable of a secondary response. The inflammatory process is a part of nonspecific immunity.

Inflammation is initiated by vasoactive amines such as histamine, serotonin, and the kinin polypeptides. Venular sphincters constrict and capillaries dilate while the kallikrein–kinin system increases vascular permeability and the adherent properties of the venular endothelium. As a result, fluid and fibrinogen leave the permeable vessels creating a fibrin network and thrombin, which contain invading bacteria. The adherent endothelial vessel walls trap phagocytes and allow their emigration into tissues. The kinins also aid in leukotaxis.

Inflammation brings serum into contact with invading microorganisms, aiding in their destruction by such nonspecific compounds as betalysin (lethal to Gram-positive organisms), and the complement cascade (lethal to Gram-negative organisms and a promoter of neutrophil phagocytosis). Complement activation leads to bacteriolysis and also the release of cleavage products with even greater activity. For example C5a and C3a, cleaved from C3, have glycoprotein components and an affinity for cell membranes (effecting vascular permeability among other things); C3a, C5a, and C567 have powerful chemotactic properties for polymorphonuclear leukocytes (PMN). Finally, inflammation results in the release and circulation of a variety of mediators with strong immunological activity, such as the prostaglandins and leukotrienes, some of which regulate lymphocyte function.

Within minutes of tissue damage or microbial invasion, PMN adhere to blood vessel walls and then emigrate into the tissue. This movement, chemotaxis, is unidirectional with no return of the cells to the circulation, and is facilitated to a large degree by complement. PMN are very active in phagocytosis, a process augmented by opsonins, which include complement and antibody. The intracellular destruction of phagocytized bacteria by PMN is facilitated by a "respiratory burst" of cellular activity, which includes:

1. increased glycolysis and lactate production;
2. a fall in the pH in phagocytic vacuoles;
3. increased O_2 consumption;
4. increased hexose monophosphate shunt activity;
5. increased NADPH and NADH oxidation;
6. increased H_2O_2 and superoxide production;
7. increased membrane lipid synthesis.

If inflammation cannot contain a local infection, invading microorganisms are carried to the regional lymph nodes via the lymphatics. Here, fixed macrophages phagocytize and often succeed in killing organisms where PMN have failed as they contain a repertoire of enzymes that is totally different. The system of fixed macrophages, contained in the spleen, lymph nodes, liver (Kupffer cells), lung (alveolar macrophages), and skin (Langerhans cells) are known collectively as the reticuloendothelial system. These cells exist in enormous numbers (approximately 200 billion are present in spleen, liver, and bone marrow) making the macrophage a central cell in the immune response. Blood-borne macrophages, or monocytes, are capable of phagocytosis and chemotaxis like PMN. Unlike PMN, however, all macrophages can undergo activation after antigen exposure, making them even more efficient in phagocytosis and intracellular killing. Lymphocytes can trigger macrophage activation by means of cellular secretions (lymphokines) and macrophages can likewise influence lymphocyte response via monokines (such as interleukin 1 (IL-1)).

The second major division of the immune reactivity is the specific immune responses of the lymphocyte. The lymphocyte population consists of two separate (more or less) response systems; the T-cell system, which is responsible for cell-mediated immunity and much of immunoregulation, and the B-cell system, which is responsible for antibody production. B and T cells arise from a common bone marrow precursor, but then mature via different pathways. Of the circulating pool of lymphocytes, approximately 80% are T cells and 12–15% are B cells. The remaining lymphocytes do not fall clearly into either category.

When stimulated by antigen, T cells undergo blast transformation then proliferate to form:

1. memory cells, which live as long as 20 years;
2. effector cells capable of a variety of responses including cytotoxicity;
3. regulatory (helper and suppressor) cells, which influence almost every aspect of the immune response, including PMN.

The regulatory activities of these specialized T cells are mediated through the elaboration of lymphokines (such as interleukin 2 and 3 (IL-2, IL-3)).

B cells are responsible for humoral immunity, the hallmark of which is the production of specific antibodies by the plasma cell progeny of activated cells. B cells can secrete five types of antibody, although it appears that only three (IgG, IgM, and IgA) participate in immunity to invading pathogens. The response of these cells and their products is specific and avid, and often carried out together with the products of the complement cascade.

LYMPHOCYTE SUBPOPULATIONS

One of the most important immunological technologies to be developed during the last decade is an ability to identify functional lymphocyte subpopulations on the basis of cell surface differentiation antigens. Monoclonal antibodies to these antigens, raised using the technique of Kohler and Milstein for the production of myeloma–lymphocyte hybrid cell lines (15), have become important tools in identifying and isolating these subpopulations. To date, several extensive series of monoclonal antibodies have been developed primarily to describe lymphocyte subsets (16–18), which are now available commercially. CD3 antibody defines an antigen present on 90–95% of all circulating, peripheral blood T cells, the CD4 antibody identifies 50–60% of T cells, and the CD8 antibody identifies 30–40% of human peripheral blood T cells. Early

Table 1.1. *Representative monoclonal antibodies that are available for the discrimination of specific lymphocyte subpopulations, their specificity, CD designations, and their commercial sources.*

CD group	Example MoAb	Specificity
CD1a	Leu 6,[a] OKT6	Cortical thymocytes, B2M-associated
CD2	OKT11	95% of thymocytes; >90% E-rosette positive lymphocytes
CD3	Leu 4; OKT3	All mature T cells and medullary thymocytes
CD4	Leu 3a,b; OKT4	Helper/inducer subset of T lymphocytes, class II MHC specific cytotoxic cells
CD5	Leu 1; OKT1;[b] T101[c]	All mature T cells and medullary thymocytes; low density on cortical thymocytes
CD8	Leu 2a,b; OKT8	Suppressor subset of T lymphocytes, class I MHC specific cytotoxic cells
CD11b	OKM1	Monocytes/phagocytic cells, some granulocytes and large granular lymphocytes
CD16	Leu 11b	Human NK cells and neutrophils
CD20	B1	All normal (IgG bearing) B lymphocytes
CD25	Tac	T cell specific activation antigen
—	NKH-1[d]	Large granular lymphocytes (LGL) including cells with natural killer activity
—	Leu 7	Medium and large lymphocytes (LGL) including cells with natural killer activity
—	Ia	90% of B lymphocytes and monocytes, 20% of null cells, activated T lymphocytes

[a]The "Leu" designated monoclonal antibodies and B1 are products of Becton-Dickinson.
[b]The OKT designated monoclonal antibodies and Ia are products of Ortho Diagnostics.
[c]T101 monoclonal antibody is a product of Hybritech.
[d]NKH monoclonal antibody is a product of Coulter Immunology.

studies showed that the CD4 antibody labeled the helper T cell subpopulation, and the CD8 antibody labeled the suppressor cells and cytotoxic T cell effectors. More recent studies, however, have shown this distinction to be an oversimplification. For example, the CD4-reactive subpopulation contains at least four separable functional subsets including both helper and suppressor cells (19), and it is now possible to distinguish CD8-positive suppressor cells from CD8-positive cytotoxic cells (20).

In spite of this imprecision in labeling lymphocyte subpopulations, monoclonal antibodies have greatly facilitated our understanding of T–T and T–B lymphocyte interactions. We now know, for example, that suppressor cells contained within the CD8-positive population require the presence of radiosensitive CD4 lymphocytes in order to function as suppressors. It is also known that CD8 suppression is directed toward CD4-positive helper cells, and not B cells, for example, in the regulation of antibody production. These probes, some of which are summarized in Table 1.1, continue to add to our knowledge of the immune system, each month bringing a multitude of scientific reports on the subject of lymphocyte subpopulations and their interactions.

Recent efforts have been made to standardize the nomenclature of monoclonal antibodies directed against leukocyte surface antigens. Monoclonals from various sources, which show similar reactivity are given a CD (cluster of differentiation) designation. Though not all antibodies have been as yet classified, this ongoing effort and the designations assigned will eventually supersede all the other systems (21). CD nomenclature is, therefore, included in the examples shown in Table 1.1.

IMMUNOLOGIC INTERACTIONS

A major emphasis in immunology today is the study of the interactions of the various components of the immune response. For example, the stimulation of suppressor-T-cell activity can result in altered B-cell activation and a reduction in antibody production. This can alter complement activation and thus PMN function. Likewise, suppressor T cells can adversely affect T-cell response and macrophage function. The products of inflammation are strong stimulators of suppressor cells and thus are very important to both specific and nonspecific immunity.

It is important to remember that, in reality, there is a close interaction of all components of immune responsiveness. Figure 1.1 illustrates some of the ways that the immune response orchestrates the destruction of pathogenic microorganisms. What affects one component of the immune response will directly or indirectly affect them all.

The monocyte/macrophage seems to be a key cell in immunological interactions as a result of its ability to secrete immunoregulatory products. Activation of the monocyte/macrophage results in changes in its secretory functions (Figure 1.2). Some of its secretory products regulate both the monocyte/macrophage and the lymphocyte population. It is now clear that PGE is one of the monocyte/macrophage products, which plays a role in regulating the immune response through its effects on various cellular populations. PGE production also appears to inhibit

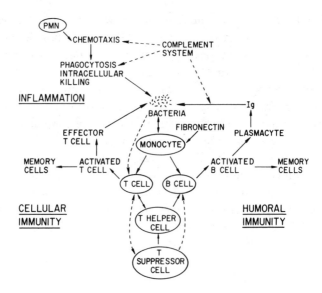

Figure 1.1. A diagrammatic summary of the interactions of the various components of the normal immune response when confronted with bacterial antigens. Each component is dependent on the efficiency of at least one other component. An effective response to a specific pathogen, however, may rely on the particular efficiency of an individual component of the immune response. PMN, polymorphonuclear leukocyte.

further activity of the monocyte/macrophage, including a down-regulation of colony stimulating factor (CSF) production. CSF is necessary to signal precursor cells to proliferate, and its secretion leads to the production of functional inflammatory cells. CSF is known to stimulate precursor stem cells for leukocytes, platelets, and erythrocytes, and to stimulate monocyte/macrophage growth, replication, and function. The production of IL-1 induces T lymphocytes to express cell surface markers and to produce lymphokines, and induces B lymphocytes to produce antibody. IL-1 also stimulates T lymphocytes to produce IL-2. IL-2 promotes the development of functional T-effector lymphocytes, and natural killer (NK) cells, and supports the growth of T cells. As a result of IL-2 stimulation, NK cells secrete gamma interferon. Interferon is also secreted by the activated monocyte/macrophage, and induces these and NK cells to express tumoricidal and bacteriolytic activity (22).

Present and future attempts to manipulate the immune response clinically, therefore, might include the administration or regulation of the production of lymphokines, interferon, or arachidonic acid metabolites.

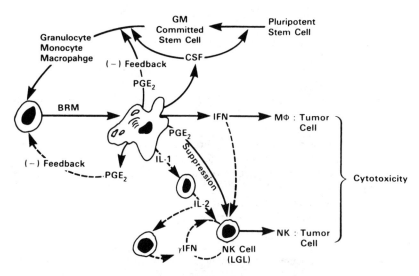

Figure 1.2. The regulation of cellular functions by secretory factors produced by the monocyte/macrophage. GM, granulocyte/monocyte; BRM, biological response modifier; PGE_2, prostaglandin E_2; IFN, interferon; $_\gamma$IFN, gamma interferon; M ϕ, macrophage; IL-1, interleukin 1; IL-2, interleukin 2; NK, natural killer cell, LGL, large granular lymphocyte. (Reprinted with permission from Chirigos MA, Schlick E, Ruffmann R: Biological response modifiers: regulation of the cellular immune system. *In*: Gruber D, Walker RI, MacVittie TJ, Conklin JJ (eds.), *The Pathophysiology of Combined Injury and Trauma*. Academic Press, New York. pp. 205–26, 1987.)

Hopes of predictably altering immune reactivity, however, must be based on a clear understanding of the influence and interaction of each of these mediators. In the case of the products of arachidonic acid metabolism, this understanding is still incomplete, as will become evident in the discussions which follow.

LITERATURE CITED

1. Kurzrok R, Lieb CC: Biochemical studies of human semen: II. The action of semen on the human uterus. *Proc. Soc. Exp. Biol. Med.* 28:268–72, 1930.
2. Goldblatt MW: A depressor substance in seminal fluid. *J. Soc. Chem. Ind.* 52:1056–61, 1933.
3. Vogt W: Uber die Beziehung des Darmstoffs zur substanz "P." *Arch. Exp. Path. Pharmakol.* 210:31–5, 1950.
4. Ambache N, Kavanagh L, Whiting J: Effect of mechanical stimulation on

10   *Prostaglandins, leukotrienes, and the immune response*

rabbits' eyes: release of active substance in anterior chamber perfusates. *J. Physiol.* (London) 176:378–408, 1965.

5. von Euler US: History and development of prostaglandins. *Gen. Pharmacol.* 14:3–6, 1983.

6. Oates JA: The 1982 Nobel Prize in Physiology or Medicine. *Science* 218:765–8, 1982.

7. Kellawey CH, Trethewie WR: The liberation of a slow reacting smooth muscle stimulating substance of anaphylaxis. *Quart. J. Exp. Physiol.* 30:121–45, 1940.

8. Harkavy J: Spasm producing substance in sputum of patients with bronchial asthma. *Arch. Inst. Med.* 45:641–6, 1930.

9. Brocklehurst WE: Slow reacting substance and related compounds. *Progr. Allergy* 6:539–58, 1962.

10. Strandberg K, Uvnas B: Purification and properties of the slow reacting substance formed in the cat paw perfused with compound 48/80. *Acta Physiol. Scand.* 82:358–74, 1971.

11. Samuelsson B, Hammarström S, Murphy RC, Borgeat P: Leukotrienes and slow reacting substance of anaphylaxis (SRS-A). *Allergy* 35:375–81, 1980.

12. Morris HR, Taylor GW, Piper PJ, Tippins JR: Structure of slow-reacting substance of anaphylaxis from guinea-pig lung. *Nature* (London) 285:204–5, 1980.

13. Lewis RA, Austen K, Drazen JM, Clark DA, Corey EJ: Slow reacting substance of anaphylaxis: identification of leukotrienes C-1 and D from human and rat sources. *Proc. Natl. Acad. Sci. USA* 78:3195–8, 1980.

14. Radmark O, Malmsten C, Samuelsson B, Clark DA, Goto G, Marfat A, Corey EJ: Leukotriene A: stereochemistry and enzymatic conversion to leukotriene B. *Biochem. Biophys. Res. Commun.* 92:954–61, 1980.

15. Kohler G, Milstein C: Continuous cultures of fused cells secreting antibody of predefined specificity. *Nature* (London) 256:495–8, 1975.

16. Engleman EG, Benike CJ, Glickman E, Evans RL: Antibodies to membrane structures that distinguish suppressor/cytotoxic and helper T lymphocyte subpopulations block the mixed leukocyte reaction in man. *J. Exp. Med.* 154:193–8, 1981.

17. Kung PC, Goldstein G, Reinherz EL, Schlossman SF: Monoclonal antibodies defining distinctive human T cell surface antigens. *Science* 206:347–9, 1979.

18. Ledbetter JA, Evans RL, Lipinski M, Cunningham-Rundles C, Good RA, Herzenberg LA: Evolutionary conservation of surface molecules that distinguish T lymphocyte helper/inducer and cytotoxic/suppressor subpopulations in mouse and man. *J. Exp. Med.* 153:310–23, 1981.

19. Chess L, Thomas Y: Human T cell differentiation. In: Twomey JJ (ed.), *The Pathophysiology of Human Immunologic Disorders.* Urban and Schwarzenberg, Baltimore pp 1–10, 1982.

20. Irigoyen OH, Rizzolo P, Thomas Y, Hemler ME, Shen HH, Friedman SM, Strominger JL, Chess L: Dissection of distinct human immunoregulatory T cell subsets by a monoclonal antibody, recognizing a cell surface antigen with wide tissue distribution. *Proc. Natl. Acad. Sci. USA* 78:3160–4, 1981.

21. Pallesen G, Plesner T: The third international workshop and conference on

human leukocyte differentiation antigens with an up-to-date overview of the CD nomenclature. *Leukemia* 1:231–4, 1987.

22. Chirigos MA, Schlick E, Ruffmann R: Biological response modifiers: regulation of the cellular immune system. In: Gruber D, Walker RI, MacVittie TJ, Conklin JJ (eds.), *The Pathophysiology of Continued Injury and Trauma.* Academic Press, New York pp 205–23, 1987.

2

Prostaglandin/leukotriene structure and chemistry: a primer

The prostaglandins and the leukotrienes are families of oxygenated fatty acids, which have been detected in virtually every mammalian tissue thus far examined. These families include some of the most potent natural substances known. Prostaglandins/leukotrienes are not stored in tissues, however, as are biogenic amines. Rather, they are formed via the activity of specific enzymes immediately prior to their release. Synthesis takes place subsequent to any cell membrane disruption, which causes the release of free fatty acids from esterified lipid sources. The precise mechanism by which the precursors of prostaglandins/leukotrienes are released is not known; however, it is agreed that they originate from the phospholipid reserves in the cell membranes. Phospholipase A is recognized as an important enzyme in this release, and recent studies with platelets have shown that at least one additional enzyme, phosphatidylinositol-specific phospholipase C, will also release membrane diacylglycerides, which can subsequently be acted upon by various lipases to produce arachidonic acid (1).

Three primary, substrate fatty acids are currently recognized as prostaglandin/leukotriene precursors. These are cis-5,8,11,14-eicosatetraenoic acid (arachidonic acid), and two related compounds, cis-8,ll,14-eicosatrienoic acid, and cis-5,8,11,14,17-eicosapentaenoic acid. Arachidonic acid is considered to be the most important of the three, so much so that the biosynthetic pathway of prostaglandin/leukotriene production is generally referred to as the arachidonic acid cascade (Figure 2.1). Arachidonic acid is metabolized, subsequent to its release, enzymatically (via a family of lipoxygenase enzymes or cyclooxygenase) or nonenzymatically (via ultraviolet light irradiation or by O_2-generating systems), and oxidation products are formed. Enzymatic oxidation of arachidonic acid via cyclooxygenase results in the formation of a series of products, which include the prostaglandins. Oxidation via the lipoxygenases yields metabolites, which include the leukotrienes. Nonenzymatic oxidation of

12

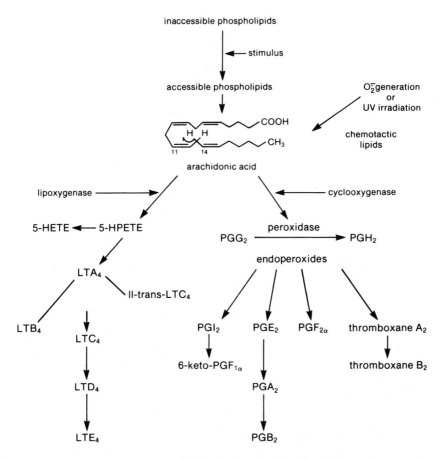

Figure 2.1. Major metabolites in the arachidonic acid cascade.

arachidonic acid is known to result in the generation of chemotactic lipids.

PROSTAGLANDIN STRUCTURE

Prostaglandins are 20-carbon (C_{20}) carboxylic acids, which contain a cyclopentane ring. For simplicity, they can be thought of as derivatives of a hypothetical C_{20} acid, given the trivial name prostanoic acid. Derivatives are sometimes termed collectively as prostanoids. The carbon atoms of the prostanoids are numbered from the carboxyl-terminal end of the molecule as indicated in Figure 2.2.

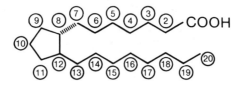

Figure 2.2. Structure of prostanoic acid.

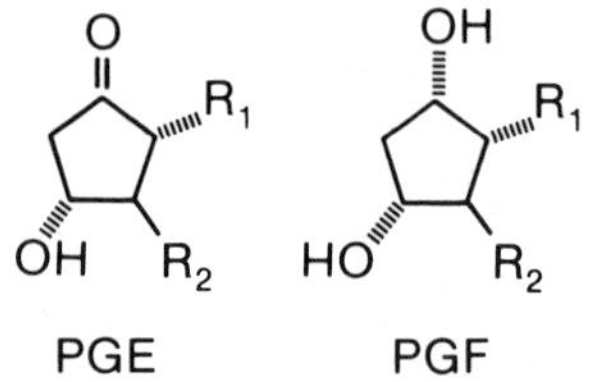

$$\text{PGE} \qquad \text{PGF}$$

Figure 2.3. Cyclopentane ring substitutions of PGE and PGF.

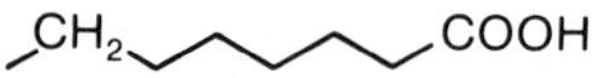

Figure 2.4. R_1 sidechain of PGE_1 and PGF_1.

There are two "primary" sets (or series) of prostaglandins: the E series and the F series. Each series has three members (prostaglandin E_1, E_2, and E_3; prostaglandin $F_{1\alpha}$, $F_{2\alpha}$, and $F_{3\alpha}$). PGEs and PGFs differ only by a single chemical substitution at the C-9 position. The E series prostaglandins possess a keto oxygen at the C-9 position of the cyclopentane ring and a hydroxyl group at the C-11, while prostaglandins of the F series have a hydroxyl group at both positions (Figure 2.3). These simple substitutions profoundly affect their biological activities.

The subscript numerals 1, 2, and 3 in each series denotes the number of double bonds in the side chains (R_1 and R_2 above) of the cyclopentane ring. In PGE_1 and $PGF_{1\alpha}$, therefore, the R_1 sidechain is saturated, as it is in the hypothetical prostanoic acid shown in Figure 2.4, while in PGE_2, PGE_3, $PGF_{2\alpha}$, and $PGF_{3\alpha}$, the R_1 sidechain contains a cis double bond at C-5 (Figure 2.5).

The R_2 sidechain in PGE_1, PGE_2, $PGF_{1\alpha}$, and $PGF_{2\alpha}$ contains a hydroxyl group at C-15 in the alpha configuration and a trans double bond at C-13. In PGE_3 and $PGF_{3\alpha}$, the R_2 sidechain contains an additional cis double bond at C-17 (Figure 2.6).

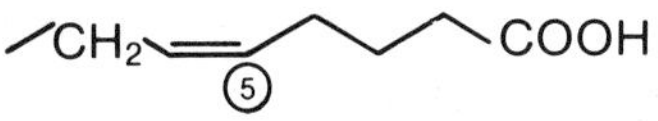

Figure 2.5. R_1 sidechain of PGE_2, PGE_3, $PGF_{2\alpha}$, and $PGF_{3\alpha}$.

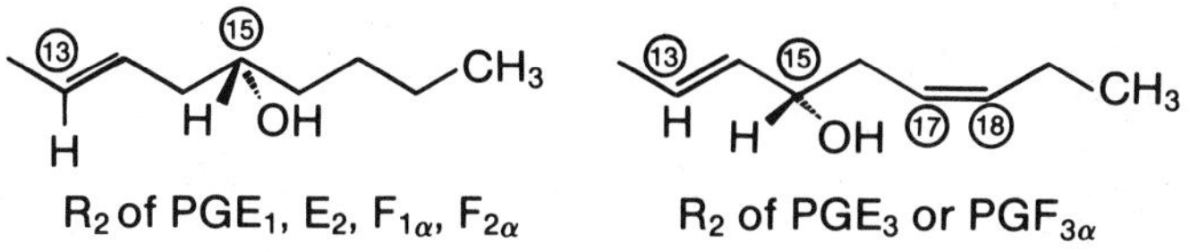

R_2 of PGE_1, E_2, $F_{1\alpha}$, $F_{2\alpha}$ | R_2 of PGE_3 or $PGF_{3\alpha}$

Figure 2.6. R_2 sidechains of PGE_1, PGE_2, $PGF_{1\alpha}$, and $PGF_{2\alpha}$, and PGE_3, and $PGF_{3\alpha}$.

The subscripts α and β after the numerical subscripts in the prostaglandin F series specify the spacial orientation of the C-9 hydroxyl group; α denotes substitution below and β denotes substitution above the plane of projection of the cyclopentane ring. All natural members of the PGF family have an α orientation at the C-9 position. PGFs with a C-9 β configuration are produced by the experimental reduction of PGE with sodium borohydride (2).

The complete structures of the six primary prostaglandins are given in Figure 2.7, which also depicts the cyclopentane ring structure of other naturally occurring "secondary" prostaglandins, most of which are generated by enzymatic or chemical reduction of the PGEs. In fact, a characteristic property of PGEs is their ability to form PGAs by acid-catalyzed dehydration, and to form PGBs under alkaline conditions.

PROSTAGLANDIN SYNTHESIS VIA CYCLOOXYGENASE

The biosynthesis of the prostaglandins is carried out in a stepwise fashion by two membrane-bound enzymes, cyclooxygenase and endoperoxide isomerase, and by the soluble enzyme peroxidase. After membrane phospholipids such as phosphatidylinositol (Figure 2.8) are released from their esterified state via phospholipase to form, for example, the stable prostaglandin precursor arachidonic acid, molecular oxygen is added to this polyunsaturated fatty acid by the enzyme cyclooxygenase at the C-9, C-11, and C-15 positions to form PGG_2. PGG_2 is quickly converted to PGH_2 by reduction of the C-15 (exo) peroxide to a hydroxyl group by the enzyme peroxidase. PGG_2 and PGH_2 are unstable, biologically active molecules, called endoperoxides, which are intermediate in the transfor-

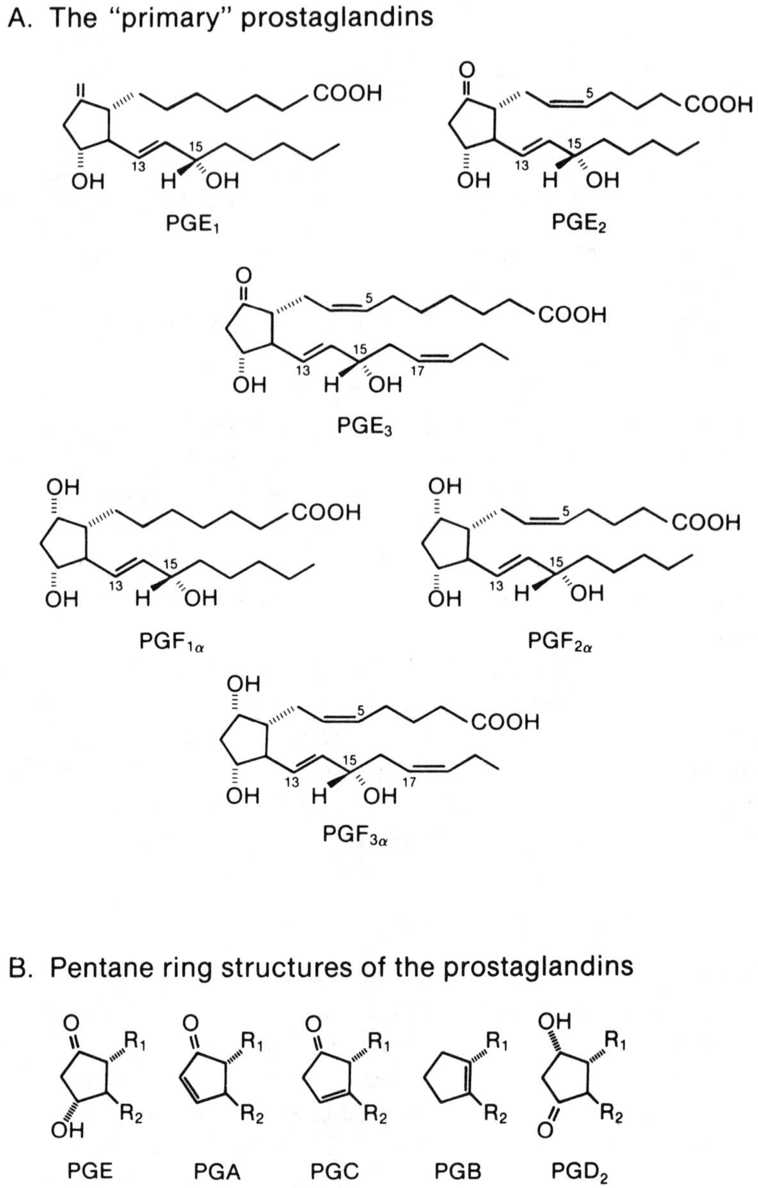

A. The "primary" prostaglandins

B. Pentane ring structures of the prostaglandins

Figure 2.7. Chemical structures of the naturally occuring PGs: (A) the "primary" PGs; (B) the pentane ring structures of the "secondary" PGs.

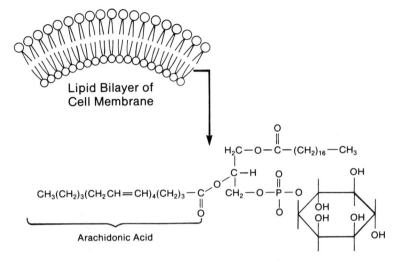

Phosphatidylinositol

Figure 2.8. Diagram of the membrane lipid bilayer, consisting of polar groups (choline, inositol, etc.) on the outside, with nonpolar, long-chain fatty acids (esterified at the C-1 and C-2 positions) on the inside. The structure of one membrane lipid, phosphatidylinositol, which serves as a source of arachidonic acid, is also shown.

mation of arachidonic acid to prostaglandins. The endoperoxides have a half-life of 4–5 min and have the ability to aggregate platelets and contract smooth muscle. Endoperoxide isomerase and peroxidase convert PGH_2 to $PGF_{2\alpha}$, PGE_2, PGI_2 (prostacyclin) and a 17-carbon cleavage product (12-L-hydroxy 5,8,10-heptadecatrienoic acid or HHT). PGG_2 can also be converted to PGE_2 through the intermediate, 15-hydroxy-peroxy-PGE_2. Prostaglandins of the A and B series are produced from the corresponding PGE by simple dehydration or, in the case of PGB, dehydration and isomerization. A soluble endoperoxide isomerase has been found in many tissues, which favors production of another prostaglandin, PGD_2, from PGE_2. Finally, the endoperoxides also serve as intermediates in the biosynthesis of thromboxane A_2 and B_2, so named because they were first isolated from thrombocytes (platelets). Thromboxane A_2 is a potent cellular regulatory agent with strong platelet-aggregating activity. Thromboxane B_2, however, is inactive. This metabolic sequence is summarized in Figure 2.9.

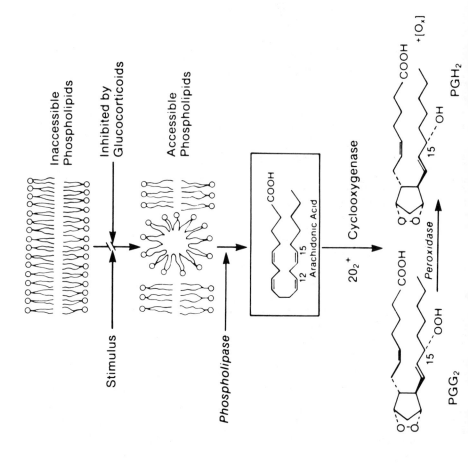

Figure 2.9. Summary of the generation of endoperoxides from arachidonic acid, and their subsequent enzymatic conversion to the PGs and thromboxanes. MDA, malondialdehyde; HHT, hydroxyheptadecatrienoic acid.

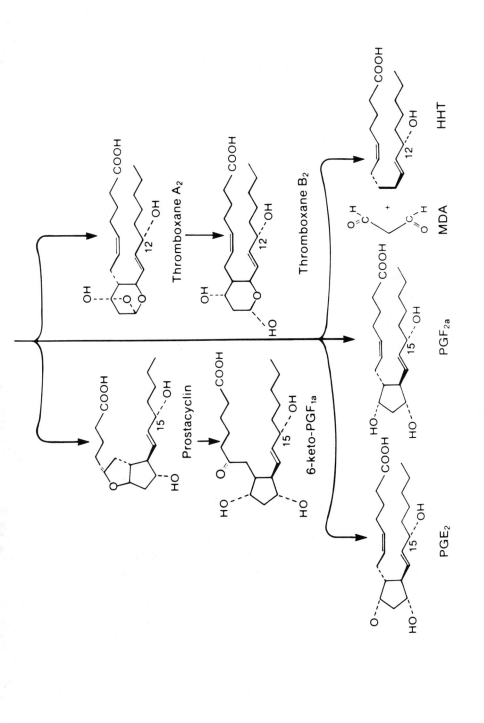

Thromboxane A₂

Thromboxane B₂

HHT

MDA

Prostacyclin

6-keto-PGF₁ₐ

PGF₂ₐ

PGE₂

The bioactive prostaglandins (PGs) are formed very quickly from their corresponding precursors, but then are quickly converted to metabolites with much weaker (or often prostaglandin inhibitory) activities. When injected into the bloodstream, PGE_2 (or $PGF_{2\alpha}$) has a half-life of less than 1 min since it is rapidly metabolized by a dehydrogenase in the lung (3). PGs produced or injected into the tissues, however, are not so quickly degraded. It has been found that the chemical addition of a methyl group at the C-15 position prevents oxidation of the 15-hydroxyl and enhances the biological activity of artificially synthesized prostaglandins (4). The half-life of PGE_2 and $PGF_{2\alpha}$ metabolites in the bloodstream is about 8 min (5). These metabolites are the 15-keto-13,14-dihydroprostaglandins, produced by the sequential action of two enzymes, prostaglandin-15-hydroxydehydrogenase and prostaglandin-13-ketoreductase. The 15-keto-13,14 dihydroxyprostaglandins are metabolized further by beta and omega oxidation to produce the urinary metabolites 7α-hydroxy-5,11,diketotetranorprostane-1,16-dioic acid (from PGE_2) and 5,7-dihydroxy-5,11-ketotetranorprostane-1,16-dioic acid (from $PGF_{2\alpha}$) (6). Total prostaglandin production for the adult human in a 24-hour period has been estimated to be 1–2 mg (7).

LEUKOTRIENE STRUCTURE

The name, leukotriene, was coined by Samuelsson to describe a family of molecules, each containing three conjugated double bonds, which are produced in leukocytes by the conversion of arachidonic acid by the calcium-dependent enzyme 5-lipoxygenase. The letter designations for the leukotrienes indicate the order of their discovery; leukotriene A (LTA), therefore, was the first described. Samuelsson had been studying the products obtained when arachidonic acid was subjected to enzymatic treatment with various lipoxygenases. He found that rabbit polymorphonuclear leukocytes (PMNs) metabolized arachidonate to a family of dihydroxy acids that showed triple spectrophotometric absorption peaks at 259, 269, and 279 nm. These triplet peaks suggested the existence of three conjugated double bonds in the molecules. Some of the PMN-derived products of arachidonate degradation had hydroxyl groups at the C-5 and 12 positions, and others were substituted at the C-5 and 6 positions. Isotope studies showed that the only hydroxyl derived from atmospheric oxygen was that in the C-5 position, while the other was derived from water. This led to the suggestion that all the metabolites had a single 5,6-epoxy-eicosatetraenoic acid precursor, which was designated leukotriene A (8). LTA_4 was subsequently isolated and purified, and the stereochemistry confirmed by total synthesis. The structures of the primary leukotrienes, LTA_4, LTB_4, LTC_4, LTD_4, and LTE_4 are given in Figure 2.10. The subscript numeral four in each indicates the presence of four double bonds in the molecule.

LEUKOTRIENE SYNTHESIS VIA LIPOXYGENASE

The enzymatic metabolism of arachidonic acid by 5-lipoxygenase results in the formation of 5-hydroperoxyeicosatetraenoic acids (5-HPETE). The common example depicted in Figure 2.10 is 5-HPETE (substituted at the C-5 position); however, substitutions catalyzed by other lipoxygenases are also possible at the C-8, 9, 11, 12, and 15 positions. PMNs are capable of forming HPETEs with substitutions at each of the first five positions concurrently (9), while other cells such as platelets appear to possess only the 12-lipoxygenase enzyme and are reported to generate only 12-HPETE (10). The HPETEs are hydroperoxides, which can be metabolized to either their analogous alcohols via peroxidase, or to leukotrienes. The alcohol derivatives of the HPETEs are reported to have strong, though nonimmunological, activities in the mammalian host. For example, 5-HETE, the alcohol formed from 5-HPETE, is reported to be a potent stimulator of the liberation of gonadotrophin-releasing hormone from cultured rat pituitary cells (11). Alternatively, 5-HPETE is metabolized by the enzyme LTA_4 synthetase to a C-5,6-transepoxide with three conjugated (7,9-trans,ll-cis) olefinic bonds, and a fourth, unconjugated double bond at C-14. This compound is LTA_4. Two enzymatic pathways have been described in the metabolism of LTA_4: (a) an LTA_4 hydrolase pathway leading to LTB_4 (5S,12R-dihydroxy 6,14-cis 8,10-trans-eicosatetraenoic acid) and (b) a glutathione s-transferase pathway, leading sequentially to the three leukotrienes, LTC_4, LTD_4, and LTE_4, which were originally described as the slow-reacting substance of anaphylaxis (SRS-A). The pathway that leads to the synthesis of SRS-A generates 5S-hydroxy, 6R-S-glutathionyl-7,9-trans,11,14-cis eicosatetraenoic acid (LTC_4) from LTA_4 by opening the epoxide with glutathione. LTC_4 is subsequently metabolized to form a 6R-S-cysteinylglycine analog (LTD_4) by the removal of glutamic acid from the peptide, and then to its 6R-S-cysteinyl analog (LTE_4) by removal of glycine.

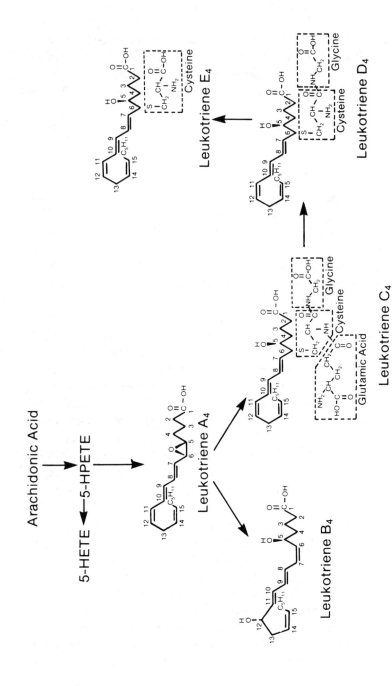

Figure 2.10. The 5-lipoxygenase pathway with the structures of the LTs.

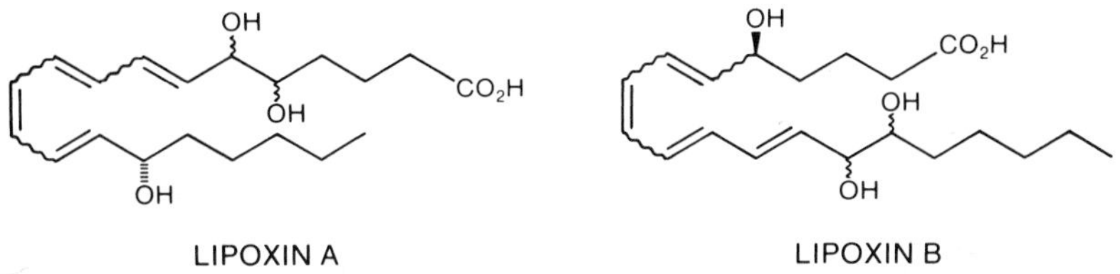

LIPOXIN A   LIPOXIN B

Figure 2.11. Proposed structures for lipoxin A and lipoxin B.

THE LIPOXINS

The initial oxygenation of arachidonic acid at the C-5 position, mediated by 5-lipoxygenase, leads to the formation of LTB_4, LTC_4, LTD_4, and LTE_4. Another set of compounds is formed via an alternative pathway in which the oxygenation of arachidonic acid takes place at the C-15 position. Mediated by the 15-lipoxygenase enzyme, initial oxygenation of arachidonic acid leads to the formation of 15-HPETE. This compound is converted to a variety of products by enzymatic and nonenzymatic processes, including 15-HETE, 8,15-diHETE, 11,12,15-triHETE, 14,15-diHETE, and 11,14,15-triHETE (12). While not yet completely studied, these compounds appear to have interesting biological properties. It has been reported, for example, that 15-HETE can inhibit leukocyte 5-lipoxygenase activity, indicating a feedback interrelationship between the 5- and 15-lipoxygenase systems (13). While studying this relationship in 1983, Samuelsson et al. isolated a new series of metabolites that were produced by human leukocytes treated with 15-HPETE (14,15). On the basis of their ultraviolet (UV) spectrum and gas chromatography-mass spectral (GC-MS) studies, two of the compounds were identified as 5,6,15-trihydroxy-7,9,11,13-eicosatetraenoic acid and 5,14,15-tri-hydroxy-7,9,11,13-eicosatetraenoic acid (5,6,15-triHETE and 5,14,15-triHETE respectively). In 1984, Serhan, Hamberg, and Samuelsson succeeded in the isolation of these metabolites and gave them the trivial names lipoxin A (LX-A) and lipoxin B (LX-B) (15). The proposed structures for these compounds are given in Figure 2.11. There are indications that the lipoxins have important biological functions, and some hints that they may possess immunological activities as well. For example, it has been reported that 5S,6S,15S,11 cis lipoxin A inhibits (NK) cell activity (16). These studies, however, are quite preliminary and, unlike the case with prostaglandins and leukotrienes, preclude any judgment concerning the immunological importance of these compounds at this time.

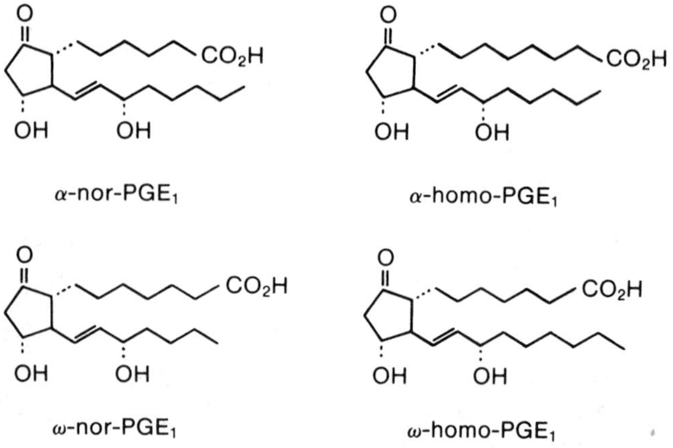

α-nor-PGE$_1$ α-homo-PGE$_1$

ω-nor-PGE$_1$ ω-homo-PGE$_1$

Figure 2.12. The structure of PGE$_1$ with sidechain variations.

STRUCTURE–ACTIVITY RELATIONSHIPS

Clues concerning the relationship of prostaglandin/leukotriene structure to their biological activities have been derived from the careful chemical and physiological analysis summarized below. Specific structural effects on immune function will be discussed in the chapters that follow. The chemical and physical analyses of the prostaglandins were initiated, for the most part, as a help in synthesizing prostaglandins with stronger or more specific activity than their native counterparts (17). As PGE$_1$ was shown very early to have a variety of biological activities (such as the stimulation of smooth muscle, gastric-acid secretion, and platelet aggregation), this prostaglandin has often been used as the standard of comparison. Whereas PGE$_2$ and PGE$_3$ were found to have a similar spectrum of activity, PGE$_3$ was found to be generally far less active, and PGE$_2$ less involved in platelet aggregation (18). Artificially varying the length of PGE sidechains also changed metabolite activity (see Figure 2.12). A-Nor- and α-homo-E prostaglandins generally have less activity than native PGE$_1$, while the omega (Ω) compounds show greatly increased potency (19). Ω-Homo-PGE$_1$ is approximately four times as active as native PGE$_1$ in platelet aggregation (18). When 8-iso-PGE$_1$ was synthesized and tested, it was found to have unaltered platelet aggregation activity; however, smooth muscle, blood pressure and lipolysis activity were greatly reduced (18). The isocompound would, therefore, be better than PGE$_1$ as a pharmacologic antithrombotic agent (20).

PGF, like PGE, generally shows high smooth muscle activity, though there are significant species variations. $PGF_{1\alpha}$, however, is approximately 30 times more active than $PGF_{1\alpha}$ in rabbit jejunum studies (21). The PGA series is reported to have hypotensive activity similar to PGE, but only weak activity in other systems (22). The 13,14-dihydroprostaglandins have the same high blood pressure and smooth muscle activities as the parent compounds; however, the 15-keto compounds are very weak (18).

Concerning the leukotrienes, it has been shown that the enzyme, 5-lipoxygenase, will accept a variety of polyunsaturated, long-chain fatty acids as substrates. Fatty acids with 19–21 carbons and 3 or more unconjugated double bonds are converted to metabolites related to 5-HETE (23). All biologically active substrates have their first double bond at C-5, while those with the first double bond at C-4, C-6, or C-8 are essentially inactive (24). Similar conclusions have been drawn from inhibition studies in which eicosa-5-y-8,11,14-trienoic acid was shown to be a powerful inhibitor of 5-lypoxygenase. In addition to conversion of LTA_4 to LTB_4, three additional dihydroxy products can be formed: two stereoisomers of $5S,6\text{-}LTB_4$ and a 12S stereoisomer of LTB_4. None of these products has biological activity. Analogues of LTB_4 with the three double bonds in the trans-cis-trans positions, the trans-trans-cis position, or all in the trans position have been shown to be inactive in chemotaxis assays (25).

It appears that the free hydroxyl function LTB_4 is necessary for its activity since acetyl LTB_4 blocks native LTB_4, and that the carboxylic acid group is also important (26). Metabolism of LTB_4 to 5,12,20-trihydroxyeicosatetraenoic acid and 5,12-dihydroxy-1,20-eicosatetraendioate, which have only partial biological activity, shows that the long alphatic chain also has importance in defining activity (23). Extensive studies have been done to determine the effect of chemical substitutions on the bioactivity of the leukotrienes, as assayed by smooth muscle activity (23). It is clear that minute changes in prostaglandin/leukotriene molecular structure translate to major alterations in biological activities, such as participation in the host immune response. These structure–immune-function relationships will be discussed in detail in the chapters which follow.

ASSAY AND MEASUREMENT METHODS

The products of arachidonic acid metabolism have extremely short half-lives, making them difficult to quantitate. In most clinical cases, prostaglandin/leukotriene levels are measured in blood plasma/serum or urine. Metabolites have also been measured, however, in joint fluids (27), cerebrospinal fluid (28), follicular fluid (29), amniotic fluid (30), inflammatory exudates (31), seminal fluid (32), gastrointestinal secretions (33), fluid from pleural or abdominal cavities (34), sweat (35), menstrual fluid (36), lymph (37), milk (38), and aqueous humor (39). There has also been an increased interest in analyzing the presence of prostaglandin/leukotrienes in tissues such as endometrium (40), gastric mucosa (41), or organs removed postmortem. Each one of these fluids and tissues presents specific problems, and, because of the instability and rapid turnover of the prostaglandins/leukotrienes, it is often hard to determine if the detection of increased metabolite levels is meaningful.

Minute amounts of the primary prostaglandins have been detected in urine, but studies have established that this is as a result of local synthesis by the kidneys (42). It is now known that prostaglandins are degraded in the body before being excreted (43). The measurement of the primary prostaglandins in urine, therefore, does not yield valid information concerning the release of these substances from specific organs or concerning total body production. Because of these considerations, studies using urine are usually confined to measurement of the major urinary metabolites: 7-dydroxy-5,11-diketotetranorprosta-1,16-dioic acid (the primary metabolite of PGE_2) and 5,7-dihydroxy-11-ketotetranorprosta-1,16-dioic acid (the primary metabolite of $PGF_{2\alpha}$).

Study of the primary prostaglandins/leukotrienes is also difficult using peripheral blood plasma/serum samples. The greatest amounts of PGE and PGF, for example, which reach the bloodstream from the tissues, occur as the corresponding 15-keto-13,14-dihydro metabolites, which have much longer half-lives than those of the respective primary prostaglandins. It is also known that primary prostaglandins are released as an artifact of blood collection from platelets or white blood cells and from the nonenzymatic cyclization of precursor fatty acids (44). This biosynthesis can be partially controlled by the addition of indomethacin and heparin to the collection tube, and by chilling and rapid centrifugation of the sample. The problem of artifactual biosynthesis during blood collection does not occur with the 15-keto-13,14-dihydro metabolites.

When the products of arachidonic acid metabolism were originally isolated, detection was accomplished by their ability to stimulate smooth muscle contraction (44). Far more precise methods of quantitation have since developed yielding today's methods of choice: radioimmunoassay (RIA), high pressure liquid chromatography (HPLC), and gas chromatography-mass spectrometry (GC-MS).

Radioimmunoassay

Radioimmunoassay (RIA) of prostaglandins was first introduced in 1970 for the measurement of PGE_1 and $PGF_{2\alpha}$ (45). Since that time, numerous assays have been developed for prostaglandins, prostaglandin metabolites, prostaglandin analogs, and thromboxanes. The advantages of RIA include high sensitivity (as low as 10 fmol), rapidity, high sample capacity, and simplicity. Disadvantages of RIA include potential nonspecificity, the need for relatively large sample volumes, and the need for a negative control or "blank" of the biological fluid under consideration.

Reliable measurements of arachidonic acid metabolites such as PGE_2 and $PGF_{2\alpha}$ in peripheral blood are very difficult, as a result of their low concentrations, the potential for blood elements such as platelets to produce the same metabolites as an artifact, and the rapid degradation and clearance of the primary prostaglandins. An especially useful RIA for $PGF_{2\alpha}$ has been developed to measure the stable 13,14-dihydro-15-keto-$PGF_{2\alpha}$ metabolite (designated PGFM). Measurement of PGE_2 is more difficult, however, since both PGE_2 and its first metabolite 13,14-dihydroketo-PGE_2 (PGEM-I) are quite unstable in aqueous environments (46). It is now known that PGEM-I degrades to 13,14-dihydro-15-keto-PGA_2 and then in the presence of albumin and high pH to 11-deoxy-13,14-dihydro-15-keto-16-cyclo-PGE_2 (PGEM-II), which is quite stable. RIA used to measure this metabolite has been shown to be a reliable indicator of PGE_2 biosynthesis (47).

The availability of relatively large quantities of synthetic leukotrienes and a knowledge of their chemistry has led to the development of specific leukotriene RIAs (48). For example, a highly sensitive RIA selective for LTB_4 has been recently described (49). Methods for the determination of leukotrienes in plasma and other biologic fluids remain poorly developed, however, because of the rapid loss of biological activity of these compounds in biological fluids, presumably caused by their interconversion and degradation (23).

High-pressure liquid chromatography

High-pressure liquid chromatography (HPLC) is particularly useful in the analysis of mixtures of the oxygenation products of arachidonic acid; however, several general problems are common with this methodology. For example, the cyclooxygenase products PGD_2, PGE_2, $PGF_{2\alpha}$, 6-oxo-$PGF_{1\alpha}$, and thromboxane B_2 (TXB_2) are so closely related in structure that they are difficult to separate. The isomeric dihydroxy products of 5-lipoxygenase activity pose similar separation difficulties, which can only be resolved via multiple-step HPLC. And finally, although LTC_4, LTD_4, and LTE_4 separate well from one another, their elution overlaps with that of other arachidonate metabolites if an appropriate mobile phase has not been selected (50). Prior to HPLC, eicosanoids are usually extracted from biological samples to minimize interference from competing or contaminating substances. This is generally accomplished using acidified aqueous solutions of diethylether or ethyl acetate (51), or columns of XAD resin (52).

Normal-phase HPLC (NP-HPLC) utilizing silica columns can be used to separate most arachidonic acid products, with the exception of the peptide leukotrienes. Recent results suggest that eluates can be monitored by UV absorbance using a mobile phase consisting of hexane:isopropanol:acetic acid (53). This method gives good separation of the monohydroxy and dihydroxy metabolites of arachidonic acid, PGD_2, PGE_2, and $PGF_{2\alpha}$, but not TXB_2 and 6-oxo-$PGF_{1\alpha}$ (which do not resolve from PGE_2) (52).

Powell has reported that reversed-phase HPLC (RP-HPLC) is the most useful approach for the separation of metabolites (50). Recoveries are generally better than those reported using silicon columns, and a variety of acceptable mobile phases are available, which allow UV monitoring. Furthermore, RP-HPLC can successfully differentiate the peptide leukotrienes. The most commonly employed mobile phases for RP-HPLC are mixtures of water and either methanol or acetonitrile, along with acetic acid or phosphoric acid. Mixtures of acetonitrile and water separate cyclooxygenase products well; however, they do not separate LTB_4 from its two 6-trans isomers (50). Mixtures of methanol and water, on the other hand, separate LTB_4 from its isomers but are not appropriate for the separation of cyclooxygenase products. Illustrative data are shown in Figure 2.13, derived from the separation of mixtures of arachidonic acid metabolites from human polymorphonuclear leukocytes and bovine lung homogenates.

Gas chromatography–mass spectrometry

Combined gas chromatography and mass spectrometry (GC-MS) has been used by many groups, as an alternative to RIA, for the study of the oxidation products of arachidonic acid. Indeed, this methodology played a key role in the original elucidation of the structure of the prostaglandins.

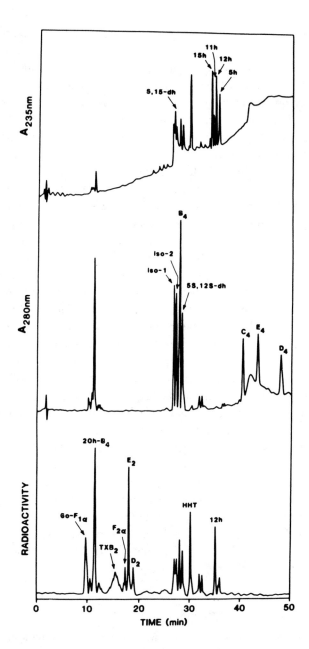

Figure 2.13. High pressure liquid chromatography (HPLC) of cyclooxygenase and lipoxygenase products using a gradient between (a) 95% solvent A (water:acetonitrile:trifluoracetic acid (TFA) 75:25:0.0008) and 5% solvent B (methanol:acetonitrile:TFA 60:50:0.002) and (b) 100% solvent B over 40 min. The flow rate was 2.0 mL/min. (Reprinted with permission from Powell WS: High-pressure liquid chromatography of arachidonic acid metabolites. *Adv. Prost. Thrombox. Leukotr. Res.* 15:53–7, 1985.)

30   *Prostaglandins, leukotrienes, and the immune response*

The relative merits of GC-MS and RIA have been extensively debated. In general, while GC-MS has detection limits which require about a 10-fold greater quantities than RIA, it is more specific for the substance of interest, and, when isotopically labeled internal standards are used, is more accurate. These advantages are countered by the high cost of sample preparation and instrumentation required for GC-MS analyses. The additional cost of the GC-MS method is greatly remediated, however, when more than one substance is to be quantified simultaneously.

While RIA is sensitive to only one substance, mass spectrometry has essentially the same sensitivity for all substances that can be volatilized. This feature makes mass spectrometry useful for analysis of an extraordinary range of substances, but also necessitates sample extraction and preparation prior to analysis. Several procedures for the extraction and enrichment of prostaglandins and thromboxanes have been previously outlined in detail. In one procedure, the sample is acidified to pH 3.5, then placed on an XAD-2 column for elution with methanol. The methanol is then evaporated and the residue is applied to an open bed reverse-phase partition column (54). The prostaglandins are eluted with a methanol/water solvent system and are then ready for GC-MS analysis. Powell has recently reported satisfactory results using a single ODS silica column (55). This procedure is especially useful because it appears to isolate all of the prostaglandins as well as their derivatives.

Prostaglandins and their oxidation products must be derivatized prior to analysis by GC-MS. In most cases this is a three-step procedure, that converts the carboxylic acid to a methyl ester, hydroxyl groups to trimethylsilyl esters, and enolizable keto groups to methyloximes. All three steps are relatively easy to perform, require a total of approximately 3 hours, and have a high yield. The final product, as a result of this procedure, is contained in the solvent N,O-bis(trimethylsilyl)acetamide (BSTFA) and may be analyzed directly by GC-MS.

LITERATURE CITED

1. Rittenhouse-Simmons S: Production of diglyceride from phosphatidylinositol in activated human platelets. *J. Clin. Invest.* 63:580–7, 1979.
2. Samuelsson B: Structures, biosynthesis, and metabolism of prostaglandins. *In*: Wakil SJ (ed.): *Lipid Metabolism*, Academic Press, New York, pp 107-53, 1970.
3. Oates JA, Roberts LD, Sweetman BJ, Maas RL, Gerkens JF, Tabber DF: Metabolism of the prostaglandins and thromboxanes. *Adv. Prostagl. Thrombox. Res.* 6:35–41, 1980.

4. Devlin, TM (ed.): *Textbook of Biochemistry with Clinical Correlations.* John Wiley & Sons, Inc., New York, pp 771–3, 1982.
5. Hamberg M, Samuelsson B: On the metabolism of prostaglandins E_1 and E_2 in man. *J. Biol. Chem.* 246:6713–21, 1971.
6. Granström E: On the metabolism of prostaglandin $F_{2\alpha}$ in female subjects. Structures of two metabolites in blood. *Eur. J. Biochem.* 27:462–9, 1972.
7. Nugteren DH: The determination of prostaglandin metabolites in human urine. *J. Biol. Chem.* 250:2808–12, 1975.
8. Borgeat P, Samuelsson B: Arachidonic acid metabolism in polymorphonuclear leukocytes: unstable intermediate information of dihydroxyacids. *Proc. Natl. Acad. Sci. USA* 76:3213–17, 1979.
9. Goetzl EJ, Sun FF: Generation of unique monohydroxy-eicosatetraenoic acids from arachidonic acid by human neutrophils. *J. Exp. Med.* 150:406–11, 1979.
10. Borgeat P, Samuelsson B: Metabolisms of arachidonic acid in polymorphonuclear leukocytes. *J. Biol. Chem.* 254:7865–9, 1979.
11. Naor Z, Vanderhoek JY, Lindner HR, Catt KJ: Arachidonic acid products as possible mediators of the action of gonadotropin-releasing hormone. *Adv. Prostagl. Thrombox. Leukotr. Res.* 12:259–64, 1983.
12. Morris J, Wishka DG: Synthesis of Lipoxin B. *Adv. Prostagl. Thrombox. Leukotr. Res.* 16:99–109, 1986.
13. Vanderhoek JY, Bryant RW, Bailey JM: Inhibition of leukotriene biosynthesis by the leukocyte product 15-hydroxy-5,8,11,13-eicosatetraenoic acid. *J. Biol. Chem.* 255:10064–6, 1980.
14. Serhan CN, Hamberg M, Samuelsson B: Trihydroxytetraenes: a novel sense of compounds formed from arachidonic acid in human leukocytes. *Biochem. Biophys. Res. Commun.* 118:943–9, 1982.
15. Serhan CN, Hamberg M, Samuelsson B: Lipoxins: novel series of biologically active compounds formed from arachidonic acid in human leukocytes. *Proc. Natl. Acad. Sci. USA* 81:5335–9, 1984.
16. Rokach J, Fitzsimmons BJ: Trihydroxytetraene metabolites of arachidonic acid: the lipoxins. *Adv. Prostagl. Thrombox. Leukotr. Res.* 16:69–81, 1986.
17. Caton MPL: Prostaglandin chemistry, structure and availability. *In:* Cuthbert MF (ed.), *The Prostaglandins: Pharmacological and Therapeutic Advances.* JB Lippincott Co., Philadelphia, pp 1–22, 1973.
18. Kloeze J: Relationship between chemical structure and platelet aggregation activity of prostaglandins. *Biochim. Biophys. Acta* 187:285–92, 1969.
19. Struijk CB, Beerthuis RK, VanDorp DA: Specificity in the enzymatic conversion of polyunsaturated fatty acids into prostaglandins. *Prostaglandins, Proceedings of the Second Nobel Symposium, Stockholm.* Almqvist & Wiksell, Stockholm, p. 51, 1967.
20. Sekhar NC, Weeks JR, Kupiecki FP: Antithrombotic activity of a new prostaglandin; 8-iso-PGE$_1$. *Circulation* 38 (suppl VI-23), 1968.
21. Horton EN, Main IHM: The relationship between the chemical structure of the prostaglandins and their biological activity. Memoirs of the Society for Endocrinology No 14, *Endogenous Substances Affecting the Human Myometrium,* p. 29, 1966.

22. Weeks JR, Sekhar NC, DuCharme DW: Relative activity of prostaglandins E_1, A_1, E_2, and A_2 on lipolysis, platelet aggregation, smooth muscle and the cardiovascular system. *J. Pharm. Pharmacol.* 21:103–8, 1969.
23. Bach MK: *The Leukotrienes; Their Structure, Actions, and Role in Diseases.* Current Concepts series, Upjohn Co., Kalamazoo, p. 12, 1983.
24. Corey EJ, Munroe JE: Irreversible inhibition of prostaglandin and leukotriene biosynthesis from arachidonic acid by 11,12 dehydro- and 5,6-dehydro-arachidonic acids, respectively. *J. Am. Chem. Soc.* 104:1752-4, 1982.
25. Sirois P, Roy S, Borgent P, Picard S, Corey EJ: Structural requirement for the action of leukotriene B_4 on the guinea pig lung: importance of double bond geometry in the 6,8,10 triene unit. *Biochem. Biophys. Res. Commun.* 99:385–90, 1981.
26. Goetzl EJ, Pickett WC: Novel structural determinants of the human neutrophil chemotactic activity of Leukotriene B. *J. Exp. Med.* 153:482–7, 1981.
27. Robinson D, McGuire M, Levine L: Prostaglandins in rheumatic diseases. *Ann N.Y. Acad. Sci.* 256:318–29, 1975.
28. Wolfe L, Mamer O: Measurement of prostaglandin $F_{2\alpha}$ levels in human cerebrospinal fluid in normal and pathological conditions. *Prostaglandins* 9:183–92, 1975.
29. Ainsworth L, Baker R, Armstrong D: Pre-ovulatory changes in follicular fluid prostaglandin F levels in swine. *Prostaglandins* 9:915–25, 1975.
30. Dray F, Frydman R: Primary prostaglandins in amniotic fluid in pregnancy and spontaneous labor. *Am. J. Obstet. Gynecol.* 126:13–19, 1976.
31. Ohuchi K, Sato H, Tsurufuji S: The content of prostaglandin E and prostaglandin F_2 alpha in the exudate of carrageenin granuloma in rats. *Biochim. Biophys. Acta.* 424:439–48, 1976.
32. Cooper I, Kelly R: The measurement of E and 19-hydroxy E prostaglandins in human seminal plasma. *Prostaglandins* 10:507–14, 1975.
33. Peskar BM, Holland A, Peskar BA: Quantitative determination of prostaglandins in human gastric juice by radioimmunoassay. *Clin. Chim. Acta* 55:21–7, 1974.
34. Velo G, Dunn C, Giroud J, Tinsit J, Willoughby D: Distribution of prostaglandins in inflammatory exudate. *J. Pathol.* 111:149–58, 1973.
35. Forström L, Goldyne M, Winkelmann R: Prostaglandin activity in human eccrine sweat. *Prostaglandins* 7:459–63, 1974.
36. Pickles V, Hall W, Best F, Smith G: Prostaglandins in endometrium and menstrual fluid from normal and dysmenorrhoeic subjects. *J. Obstet. Gynecol. Br. Commonw.* 72:185–92, 1965.
37. Anggard E, Johnson C: Efflux of prostaglandins in lymph from scalded tissue. *Acta. Physiol. Scand.* 81:440–7, 1971.
38. Mams J: The excretion of prostaglandin F_2 alpha in milk of cows. *Prostaglandins* 9:463–74, 1975.
39. Paterson C, Pfister R: Prostaglandin-like activity in the aqueous humor following alkali burns. *Invest. Ophthalmol.* 14:177–83, 1975.

40. Green K, Hagenfeldt K: Prostaglandins in the human endometrium. *Am. J. Obstet. Gynecol.* 122:611–14, 1975.
41. Peskar BM, Peskar BA: On the metabolism of prostaglandins by human gastric fundus mucosa. *Biochem. Biophys. Acta* 424:530–8, 1976.
42. Frolich J, Wilson T, Sweetman B, Smigel M, Nies A, Carr K, Watson J, Oates J: Urinary prostaglandins: Identification and origins. *J. Clin. Invest.* 55:763–70, 1975.
43. Samuelsson B, Granström E, Green K, Hamberg M, Hammarström: Prostaglandins. *Am. Rev. Biochem.* 44:669–95, 1976.
44. Granström E, Samuelsson B: Quantitative measurement of prostaglandins and thromboxanes: general considerations. *Adv. Prostagl. Thromb. Res.* 5:1–13, 1978.
45. Levine L, Van Vunakis H: Antigenic activity of prostaglandins. *Biochem. Biophys. Res. Commun.* 41:1171–7, 1970.
46. Mitchell MD, Sors H, Flint APF: Instability of 13,14-dihydro-15-keto-prostaglandin E_2. Lancet i:558, 1977.
47. Mitchell MD, Ehenhack K, Kraener DL, Cox K, Cutrer S, Strickland DM: A sensitive radioimmunoassay for 11-deoxy- 13,14-dihydro-15-keto-11,16-cyclo-prostaglandin E_2: application as an index of prostaglandin E_2 biosynthesis during human pregnancy and parturition. *Prostagl. Leukotr. Med.* 9:549–57, 1982.
48. Levine L, Morgan RA, Lewis RA, Austen KF, Clark DA, Marfat A, Corey EJ: Radioimmunoassay of the leukotrienes of slow reacting substance of anaphylaxis. *Proc. Natl. Acad. Sci. USA* 78:7692–6, 1981.
49. Salmon JA, Summons PM, Palmer RMJ: A radioimmunoassay for leukotriene B_4. *Prostaglandins* 24:225–35, 1982.
50. Powell WS: High pressure liquid chromatography of arachidonic acid metabolites. *Adv. Prostagl. Thrombox. Leukotr. Res.* 15:53–7, 1985.
51. Powell WS: High pressure liquid chromatography of eicosanoids. *In:* Lands WEM (ed.), *Biochemistry of Arachidonic Acid Metabolism.* Martinus Nijhoff, Boston, pp. 375–403, 1985.
52. Green K, Hamberg M, Samuelsson B, Frolich J: Extraction and chromatographic procedures for purification of prostaglandins, thromboxanes, prostacyclin, and their metabolites. *Adv. Prost. Thrombox. Res.* 5:15–38, 1978.
53. Borgeat P, Picard S, Vallerand P, Sirois P: Transformation of arachidonic acid in leukocytes. Isolation and structural analysis of a novel dihydroxy derivative. *Prost. Med.* 6:557–70, 1981.
54. Green K, Hamberg M, Samuelsson B, Frolick JC: Measurement of prostaglandins, thromboxanes, prostacyclin, and their metabolites by gas-liquid chromatography-mass spectrometry. *Adv. Prost. Thrombox. Res.* 5:39–94, 1978.
55. Powell WS: Rapid extraction of oxygenated metabolites of arachidonic acid from biological samples using octadecylsilylsilica. *Prostaglandins* 20:947–57, 1980.

3

Monocytes and macrophages

The monocyte/macrophage has a variety of immunological functions including presentation of antigen to lymphocytes, production of immunologically active mediators such as interleukin 1(IL-1), and regulation of cell–cell cooperation and activity. This cell is essential for the development of almost all normal immunological responses. Activated macrophages can also act as cytotoxic effector cells that are especially critical to antimicrobial and antitumor defenses (1).

Monocytes, macrophages, and macrophagelike cells produce a number of mediators critical to immune responsiveness. For example, IL-1 is secreted by macrophages, macrophagelike cells (such as Langerhans cells of the skin), and by astrocytes, keratinocytes, fibroblasts, and natural killer (NK) cells (2). Collectively, these mediator-producing cells are called *accessory cells* (3). Accessory-cell IL-1 is directly involved in the activation of helper-T lymphocytes, and the stimulation of a variety of other cell types to produce "secondary mediators" with far-reaching influence. As summarized in Figure 3.1, these secondary mediators include prostaglandins and leukotrienes.

THE EFFECT OF PROSTAGLANDINS ON MACROPHAGE FUNCTION

The products of arachidonic acid metabolism are important regulators of macrophage activity and macrophage–lymphocyte cooperation. Many studies have been devoted to examining the effects of prostaglandin E on macrophages. In these studies, PGE has been shown to have a profound inhibitory effect on the release of lysosomal hydrolase in human macrophages, as well as on macrophage locomotion, shape change, and phagocytosis (4). These responses are associated with an increase in cyclic AMP levels, although, in some cases, functional changes seem to result from direct PGE–membrane interactions (5). PGI_2 has been demon-

34

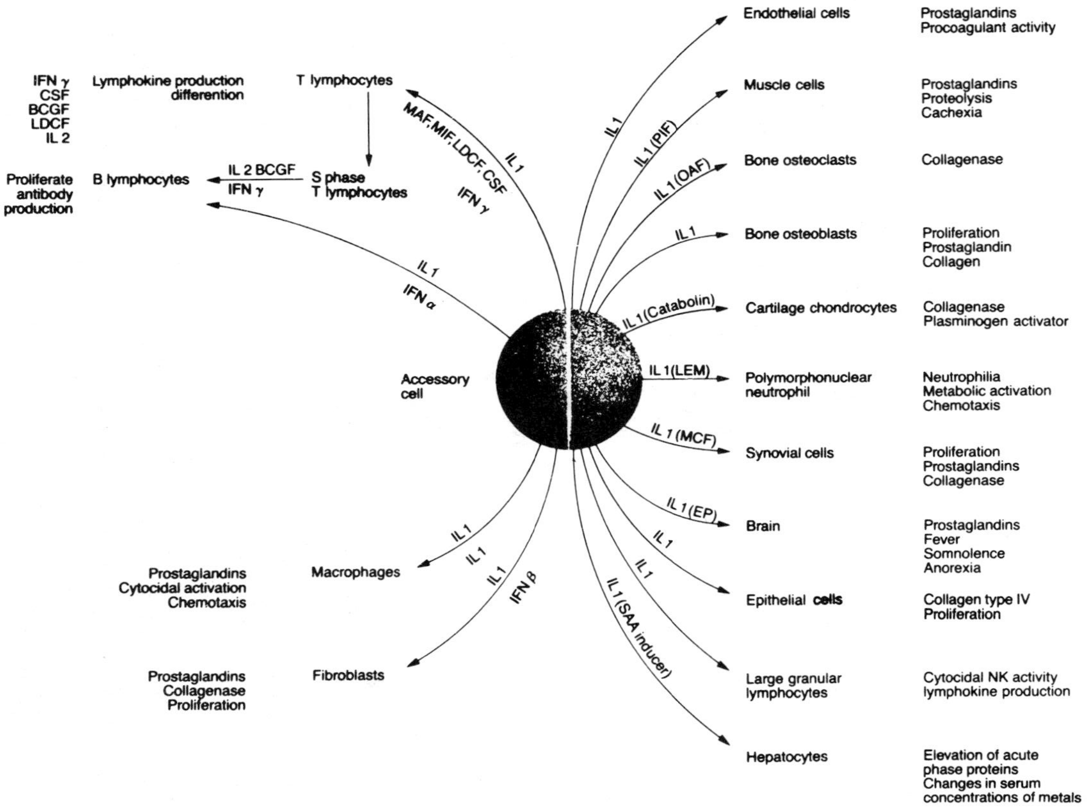

Figure 3.1. The effects of accessory-cell IL-1 on various cells and tissues. Accessory-cell-derived IL-1 enhances the growth and function of many target cells. Conversely, cytokines produced by these cells in response to IL-1 can modulate accessory cell functions. PIF, proteolysis inducing factor; OAF, osteoclast activating factor; LEM, leukocytic endogenous mediator; MCF, mononuclear cell factor; EP, endogenous pyrogen; SSA, amyloid A; CSF, colony stimulating factor; BCGF, B cell growth factor; LDCF, lymphocyte-derived chemotactic factor; MAF, macrophage activating factor; MIF, macrophage inhibitory factor. (Reprinted with permission from Oppenheim JJ: *Interleukins and Interferons in Inflammation*. Current Concepts series. Upjohn Co., Kalamazoo, 1986.)

Endothelial cells — Prostaglandins, Procoagulant activity

Muscle cells — Prostaglandins, Proteolysis, Cachexia

Bone osteoclasts — Collagenase

Bone osteoblasts — Proliferation, Prostaglandin, Collagen

Cartilage chondrocytes — Collagenase, Plasminogen activator

Polymorphonuclear neutrophil — Neutrophilia, Metabolic activation, Chemotaxis

Synovial cells — Proliferation, Prostaglandins, Collagenase

Brain — Prostaglandins, Fever, Somnolence, Anorexia

Epithelial cells — Collagen type IV, Proliferation

Large granular lymphocytes — Cytocidal NK activity, lymphokine production

Hepatocytes — Elevation of acute phase proteins, Changes in serum concentrations of metals

T lymphocytes — Lymphokine production, differentiation; IFN γ, CSF, BCGF, LDCF, IL 2

B lymphocytes → S phase T lymphocytes — IL 2 BCGF, IFN γ; Proliferate, antibody production

Accessory cell

Macrophages — Prostaglandins, Cytocidal activation, Chemotaxis

Fibroblasts — Prostaglandins, Collagenase, Proliferation

Arrow labels: IL 1, IL 1 (PIF), IL 1 (OAF), IL 1 (Catabolin), IL 1 (LEM), IL 1 (MCF), IL 1 (EP), IL 1, IL 1 (SAA inducer), IFN β, IL 1, IFN α, MAF, MIF, LDCF, CSF, IFN γ strated to produce a similar increase in cyclic AMP levels and functional changes in mouse peritoneal macrophages (6).

Studies *in vivo* have suggested that the intensity of the effect of prostaglandins on macrophages is related to the level of macrophage differentiation. When macrophages, isolated from carrageenan-induced granulomas, are exposed to PGE_2, they respond with vigorous production of intracellular cyclic AMP. The magnitude of the cyclic AMP increase is clearly dependent upon the stage of inflammatory development reached by the granuloma tissue from which the cells are removed (7). Although PGE_2 and PGI_2 could be demonstrated to have similar capacities to elevate cyclic AMP levels in peritoneal macrophages, PGI_2 was not at all effective in increasing cyclic AMP levels in granuloma macrophages highly responsive to PGE_2 (8).

Macrophages activated by interferon and bacterial lipopolysaccharide (LPS) to express tumoricidal activity are inhibited by high doses of PGE_2 (9,10). PGE_2 also reduces the cytotoxic effects of Bacillus Calmette-Guerin (BCG)-activated macrophages (11). Phagocytosis, however, may be either enhanced or inhibited by PGE_2, depending upon the state of cell differentiation and PGE_2 concentration (12,13). Likewise, low concentrations of PGE_2 (10^{-9} M) were found to suppress macrophage adherence and spreading, while higher concentrations (10^{-6} M) were found to enhance migration (13).

The macrophage appears to have specific binding sites (receptors) for PGE_2 through which most of these effects are mediated. In studies of the specific binding of $[^3H]PGE_2$ to mature macrophages, Meerpohl and Bauknecht found that binding reached a plateau after 90 min incubation (14). Scatchard-plot analysis (Figure 3.2) revealed a homogeneous population of binding sites with a dissociation constant (K_d) of 2×10^{-9} M. The binding capacity reached 7–9 fmol/10^7 cells, independent of the incubation temperature.

EICOSANOIDS AND ANTIGEN PRESENTATION

Snyder et al. studied the effect of prostaglandins on the expression of I-region-associated (Ia) antigens by macrophages (15). Only a percentage of phagocytes bear Ia molecules on their surface, and it is these cells which are responsible for macrophage–lymphocyte interactions, including induced proliferation of T cells. The importance of Ia expression can be summarized as follows:

1. T cells will not grow in antigen-stimulated cultures unless macrophages are present;

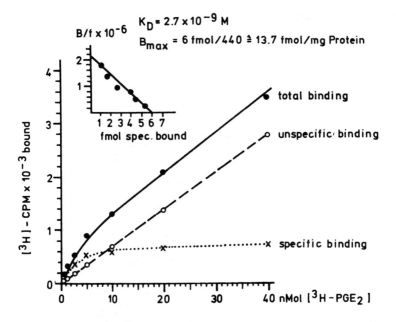

Figure 3.2. Scatchard plot analysis of PGE_2 binding to murine bone marrow macrophages. Cultures contained 5×10^6 cells in Tris-HCL buffer at pH 7.0. Incubation in $[^3H]PGE_2$ was carried out at 37 ° for 45 min. (Reprinted with permission from Meerpohl HG, Bauknecht T: Role of prostaglandins on the regulation of macrophage proliferation and cytotoxic functions. *Prostaglandins*. 31:961–71, 1986.)

2. macrophages are required as antigen-presenting cells;
3. the I region regulates interaction of T cells with macrophages;
4. macrophages which interact with T cells must of necessity bear Ia antigen on their surface;
5. I-region involvement in macrophage–lymphocyte interaction applies to all antigens; and
6. the interaction between the T cell and the appropriate macrophage selects and maintains the life of the antigen- Ia-reactive T-cell clone (16).

Ia expression, therefore, is essential for macrophages to function as antigen presenting cells during the induction of an immune response.

The synthesis and membrane expression of Ia by mature macrophages is stimulated in the presence of activated T lymphocytes (17), and inhibited by the presence of young, replicating macrophages (18). Snyder explains this balanced regulation of Ia expression in terms of prostaglandin E production, which strongly inhibits Ia antigens on macrophages, and on the presence of thromboxane B_2 (TXB_2), which antagonizes

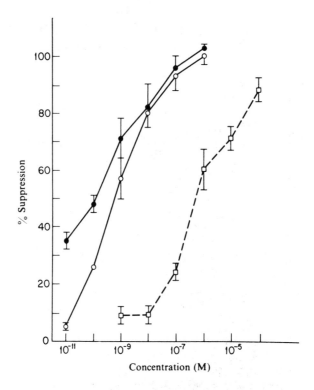

Figure 3.3. Suppression of macrophage Ia expression by PGE_1 (closed circles), PGE_2 (open circles), and dibutyryl cyclic AMP (open squares) in peptone-elicited murine peritoneal macrophages. (Reprinted with permission from Snyder DS, Beller DI, Unanue ER: Prostaglandins modulate macrophage Ia expression. *Nature* (London) 299:163–5, 1982.)

PGE effects (15). Suppression of Ia expression by macrophages in cultures supplemented with PGE_1, PGE_2, or dibutyryl cyclic AMP is shown in Figure 3.3. Significant suppression occurs in the presence of all three supplements. Dibutyryl cyclic AMP inhibited Ia expression with an I_{50} of 9×10^{-7} M, and both PGE_1 and PGE_2 with an I_{50} of approximately 10^{-10} M. The effect of TXB_2 on PGE_2-induced suppression of macrophage Ia antigen expression is shown in Figure 3.4. The presence of TXB_2 prevents Ia suppression by PGE_2; however, these effects were evident only in the presence of T-cell stimulation. Preliminary experiments also showed that three monohydroxyeicosatetraenoic acids (5-HETE, 11-HETE, and 15-HETE) were also effective in increasing the number of Ia-positive macrophages (15).

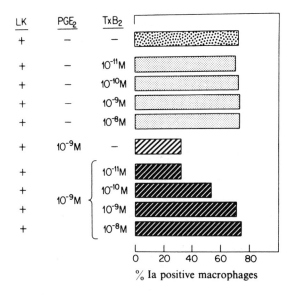

Figure 3.4. Demonstration of the antagonistic effects of TXB_2 and PGE_2. Murine peritoneal macrophages were cultured in the presence of 2% T cell lymphokine (LK), plus PGE_2, TXB_2, or both. Cells were pulsed with heat-killed *Listeria* organisms on day 4, then fixed and stained for Ia on day 5. (Reprinted with permission from Snyder DS, Beller DI, Unanue ER: Prostaglandins modulate macrophage Ia expression. *Nature* (London) 299:163–5, 1982.)

MACROPHAGE PRODUCTION OF EICOSANOIDS

Not only are macrophages affected by the presence of cyclooxygenase and lipoxygenase products, they are also clearly a major source of immunologically active eicosanoids (19–23). Humes et al. showed that murine macrophages synthesized and released prostaglandins in response to inflammatory stimuli (24), and Brume et al. showed that PG release from macrophages was associated with the formation of large intracellular vacuoles (25).

The identification of the monocyte as the primary human peripheral blood mononuclear cell responsible for PG production (specifically PGE) was largely the result of work by Bankhurst et al. (26), who demonstrated cytoplasmic PGE in monocytes via immunofluorescence (Table 3.1). Human monocytes/macrophages also metabolize arachidonic acid to prostacyclin and thromboxane, which can be measured by the pres-

Table 3.1. *The nature of the mononuclear cell containing cytoplasmic PGE as measured by indirect immunofluorescence.*

	Unfractionated cell suspensions[a]	Glass-nonadherent cells	Glass-adherent cells
Percent cytoplasmic PGE by immunofluorescence	13.0 ± 2.0[b]	1.0 ± 0.5	48 ± 3
Percent monocytes by histo-chemical criteria	12.5 ± 0.4	0.5 ± 0.3	62 ± 5
Percent monocytes by Wright's stain morphology	15.0 ± 0.7	1.5 ± 0.3	62 ± 8
Percent cells with ingested latex particles	12.5 ± 2.4	0.5 ± 0.3	52 ± 4
Percent cells with latex parti-cles that also have cytoplasmic PGE	95.0 ± 3.0	—	100

[a] Cell suspensions consisted of fresh, unstimulated cells in each case.
[b] The numeric expression represents the mean ± SE of two experiments.

Source: Reprinted with permission from Bankhurst AD, Hastain E, Goodwin JS, Peake GT: The nature of the prostaglandin-producing mononuclear cell in human peripheral blood. *J. Lab. Clin. Med.* 97:179–86, 1981.

ence of the stable metabolites 6-keto-PGF$_{1\alpha}$ and TXB$_2$ in culture supernatants (27).

It was also clear from the studies of Bankhurst and others, that subpopulations of monocytes exist, which differ in their ability to produce prostaglandin. Just as macrophage response to the presence of eicosanoids is related to the state of cell differentiation, the "experience" of the macrophage also has an important influence on PG synthesis by these cells. Humes et al., for example, studied PG synthesis by murine peritoneal macrophages following stimulation of mice with either thioglycollate broth or *Corynebacterium parvum*, and found that approximately equal quantities could be detected under basal conditions (24). However, while macrophages from unstimulated animals responded to zymosan challenge with a 20–30-fold increase in PG production, the response of stimulated macrophages was much smaller (approximately 5-fold). Normal (unstimulated) macrophages also reacted to the presence of antigen–antibody complexes or the reducing agent, phorbol myristate acetate (PMA), with marked increases in PG synthesis (as much as 60-fold), as early as 30 min postchallenge, and continuing for at least a day in culture. Phagocytosis of latex beads, however, did not produce the same effect (24).

Macrophage PG synthesis is greatly increased in the presence of calcium ionophore A23187 (Figure 3.5). Ionophorous compounds, such as A23187, have been employed in many studies of cellular activation such as lymphocyte blastogenesis, macrophage activation, production of eosinophil chemotactic factor, and the induction of histamine release by basophils (reviewed in 28). While it was initially assumed that ionophores functioned by increasing intracellular Ca^{2+}, it has been shown that ionophore activation of lymphocytes can take place in the absence of Ca^{2+} translocation (29). It is now known that the ionophore A23187 induces membrane changes such as increases in phospholipid and phosphatidylinositol turnover (29,30). In their study of rat peritoneal macrophages, Gemsa et al. showed that these membrane changes included the release of PGE and, to a lesser extent, PGF (28). Therefore, if we accept these results as examples of events which take place during macrophage activation, PG synthesis as part of a feedback regulation is suggested to be a normal part of macrophage response.

Macrophages which produce PGE bring about suppression of an immune response by:

1. PGE$_2$ induction of suppressor T cells;
2. direct interaction with lymphocytes in culture; and
3. acting as mediators or carriers of T-cell suppressor factors (24).

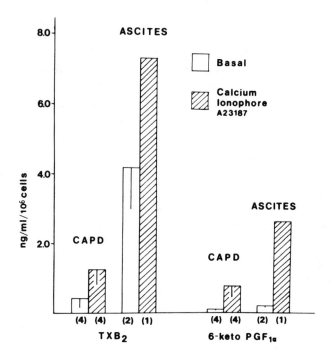

Figure 3.5. Basal release of TXB_2 and 6-keto-$PGF_{1\alpha}$ compared to the release of these metabolites following incubation with calcium ionophore A23187 (10^{-6} M to 0.2×10^{-6} M) in human macrophages obtained by continuous ambulatory peritoneal dialysis (CAPD) and ascites macrophages. The stimulated release of both TXB_2 and 6-keto-$PGF_{1\alpha}$ is significantly higher than basal release in all cases. (Reprinted with permission from Foegh M, Maddox YT, Winchester J, Rakowski T, Schreiner G, Ramwell PW: Prostacyclin and thromboxane release from human peritoneal macrophages. *Adv. Prostagl. Thrombox. Leukotr. Res.* 12:45–9, 1983.)

Macrophages which function in this capacity are sometimes called *suppressor macrophages*. Their suppressive activity can be reversed in culture by the addition of cyclooxygenase inhibitors such as indomethacin (24).

Stimulated human macrophages are also able to synthesize and release 5-lipoxygenase products such as LTB_4 and LTC_4 (31,32), and it is thought that release of these compounds acts as a counterbalance to the effects of PGE_2 synthesis. The immunological consequences of macrophage release of both cyclooxygenase and lipoxygenase products are summarized in Figure 3.6, which describes the positive–negative feed-

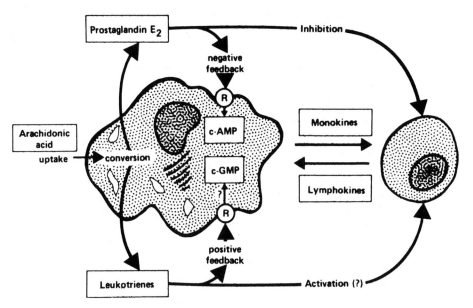

Figure 3.6. PG- and LT-mediated feedback mechanisms in macrophage–lymphocyte interaction. The macrophage, shown on the left, converts arachidonic acid substrate to immunologically active PGE_2 and leukotrienes (LTB_4). These metabolites effect both macrophage metabolism, and also the response of lymphocytes, shown on the right. (Reprinted with permission from Bonta IL, Parnham MJ: Immunomodulatory-antiinflammatory functions of E-type prostaglandins. Minireview with emphasis on macrophage-mediated effects. *Int. J. Pharmacol.* 4:103–9, 1982.)

back regulation of macrophage–lymphocyte cooperation via eicosanoid synthesis by activated macrophages. Basically, PGE_2 production leads to both inhibition of lymphocyte responses and inhibition of the PGE_2-producing macrophage, through interaction with cell surface receptors and subsequent stimulation of intracellular cyclic AMP. Leukotriene production by macrophages, on the other hand, appears to stimulate lymphocyte reactivity (4,33). Leukotrienes, however, have been reported to promote the release of macrophage PGE_2, leading to the concept of self-limited macrophage activity via shifts in eicosanoid production (34).

T lymphocytes appear to be able to induce and to regulate eicosanoid metabolism by the macrophage by supplying the necessary arachidonic acid substrate. This has been tested in models of TXB_2 production (35) and the production of PGE_2 (36) by macrophages (Table 3.2). Conversely, receptors for PGE are present on T cells, and are found particularly in

Table 3.2. *Immunoreactive TXB$_2$ and PGE$_2$ generation by T cells alone, or T cells cultured with PHA, monocytes (Mϕ), or monocytes plus PHA.*

	(ng/ml ± SE, N = 2)	
Additions	iTXB$_2$	iPGE$_2$
Medium alone	1.64 ± 0.3[a]	0.19 ± 0.2
T cells (5 × 10⁶)	1.31 ± 0.7[a]	<0.01
T cells + 5 μg/ml PHA	1.21 ± 0.7[a]	0.10 ± 0.1
Mϕ (1 × 10⁶)	17.3 ± 1.4	2.5 ± 0.6
T cells (5 × 10⁶) + Mϕ (1 × 10⁶)	22.5 ± 2.7[b]	3.6 ± 0.3
T cells + Mϕ + 5 μg/ml PHA	43.8 ± 1.8[b]	6.2 ± 0.6

[a] There were no significant differences between cultures of T cells alone, T cells + PHA, or macrophages alone.
[b] Cultures of macrophages + T cells + PHA were significantly different (p<0.05) from unstimulated T cells + macrophages for TXB$_2$ levels only.

Source: Reprinted with permission from Goldyne ME, Stobo JD: T-cell-derived arachidonic acid and eicosanoid synthesis by macrophages. *Adv. Prostagl. Thrombox. Leukotr. Res.* 12:39–43, 1983.

a subpopulation of T lymphocytes that is capable of binding the Fc fragment of IgG (37). This subpopulation, designated Tγ cells, are lymphocytes which can act as suppressor cells in the humoral response *in vitro* induced by pokeweed mitogen (PWM) (38). Kaszubowski and Goodwin showed that the addition of either exogenous PGE$_2$ or the supernatants from macrophage cultures *in vitro* induced increased Fcγ receptor expression by human T lymphocytes (39). These results are shown in Figure 3.7.

IG AND LEUKOTRIENE PRODUCTION

Human monocytes bear receptors for the Fc portion of immunoglobulins of various classes (40) and it has been shown that expression of these receptors varies with disease (41,42). Ferreri et al. found that interaction of IgG, IgA, and IgE (but not IgM) with monocyte surface receptors correlated with the release of LTB$_4$, LTC$_4$, and PGE$_2$ from these cells (43). Representative results are shown in Figure 3.8. Maximal leukotriene (LT) release was observed after 30 min; however PGE$_2$ production increased for up to 2.5 h. Production of LTB$_4$ and LTC$_4$ was completely inhibited by the removal of calcium from the medium, or

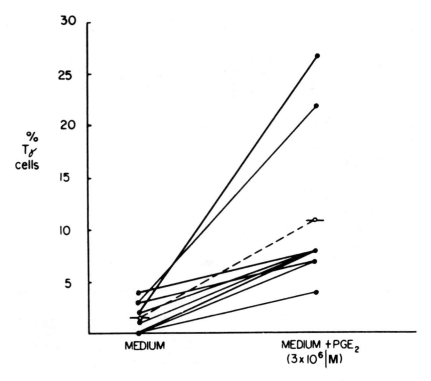

Figure 3.7. Percentage of T_γ cells within the "non-T_γ" population after 18-h incubation with or without 3×10^6 M PGE_2. In 9 experiments, the mean percentage ± standard error of T_γ cells among the "non-T_γ" fraction incubated in medium alone was 1.5 ± 1.4%. This increased to 11 ± 7.8% T_γ cells when PGE was added ($p < 0.001$, by paired t-test). (Reprinted with permission from Kaszubowski PA, Goodwin JS: Monocyte-produced prostaglandin induces Fc_γ receptor expression on human T cells. *Cell. Immunol.* 68:343–8, 1982.)

preincubation of the cells with nordihydroguaiaretic acid (NDGA), a leukotriene inhibitor, and production of PGE_2 was 80% inhibited by the addition of indomethacin.

Available data suggest that IgG, IgA, and IgE immune complexes induce metabolism of monocyte arachidonic acid by cross-linking Fc receptors. Important to the disease process, eicosanoid production does not require phagocytosis, suggesting that the exposure of monocytes to immobilized immune complexes (deposited in blood vessel walls, for example, or on glomerular basement membrane) can initiate arachidonic acid metabolism.

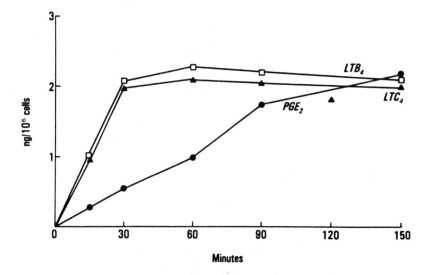

Figure 3.8. Courses of LTC_4, LTB_4, and PGE_2 release from human monocytes treated with aggregated IgG. The metabolites were quantitated using RIA. This figure represents data generated in 6 experiments. (Reprinted with permission from Ferreri NR, Howland WC, Spiegelberg HL: Release of leukotrienes C_4 and B_4 and prostaglandin E_2 from human monocytes stimulated with aggregated IgG, IgA, and IgE. *J. Immunol.* 136:4188–93, 1986.)

While these experiments were performed with human peripheral blood monocytes, it is known that human peritoneal macrophages also synthesize LTB_4 and LTC_4 (44). IgE has also been implicated in the induction of LTC_4 release from mouse and rat macrophages in a prolonged fashion (45,46). This observation may be important to our understanding of atopic (allergic) disease, which is discussed in Chapter 10.

INTERFERON AND PROSTAGLANDINS

It has been proposed that interferon and PGE have an inverse relationship with regard to macrophage activity (47). Both interferons and PGs are released during the inflammatory response, and probably act in counterbalance to each other in the inflammatory process. PGE can suppress cellular functions that have been shown to be enhanced by interferon, including macrophage activation, NK cell activity, and lymphocyte blastogenesis (48). Paradoxically, in hyporeactive animals, the addition of PG has been shown to restore the macrophage interferon response, suggesting a positive-feedback function (48). These reports again support the hypothesis that the state of differentiation of the macrophage is a determinative factor in the response elicited by PGE.

The production of LTC_4 by mouse resident peritoneal macrophages, stimulated with zymosan or the calcium ionophore A23187, is markedly inhibited when the cells are preexposed to IFN_α, IFN_β, or IFN_γ (49). This inhibition is due to both a decrease in available cellular arachidonic acid as a result of interferon treatment, and an inhibition in LTA_4 synthesis. Boraschi et al. interpreted these results to indicate possible interferon regulation of local anaphylaxis and inflammatory reactions (49).

STIMULATORY EFFECTS OF PROSTAGLANDIN SYNTHESIS

Not all the effects of macrophage-produced PGs are inhibitory, although the foregoing discussion of suppression and suppressor cells would imply that this is the case. Macrophage-produced PG stimulates the production of several important mediators including osteoclast-activating factor by T cells (49) and the release of collagenase from macrophage/monocytes, and may play a positive role in epidermal-cell growth.

It has been known for some time that endotoxin (and various lymphokines) induce collagenase production by macrophages. This effect has been found to be linked to PGE_2-mediated increases in cyclic AMP (50). Such collagenase production can be blocked by indomethacin and restored by the addition of exogenous PGE_2.

Conversely, it appears that collagen (types II and III) can stimulate PGE_2 production by monocytes. Dayer et al. studied the response of mononuclear cells from six patients with rheumatoid arthritis and six normal subjects to human types I, II, and III collagen (51). Cells isolated from all 12 subjects synthesized PGE_2, as well as mononuclear cell factor (MCF), in response to collagen type II and III, while collagen type I had little effect (Figure 3.9). Cells from patients with rheumatoid arthritis had a similar response, but displayed important differences in other areas such as production of leukocyte inhibitory factor (LIF) (51). The results of this study and others (52) suggest that the presence of major structural macromolecules (such as collagen) in the joint space, modulate the production of immunological mediators which may contribute to the immunopathology of rheumatoid disease. This possibility is discussed in greater detail in Chapter 8.

Finally, it appears that PGE_2 synthesis may play a role in the promotion of epithelial proliferation and wound healing. It is possible that

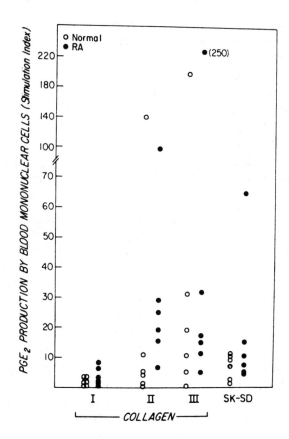

Figure 3.9. Stimulation of PGE_2 release into supernatants of 24 h cultures of blood mononuclear cells, from 6 normal subjects and 6 patients with rheumatoid arthritis (RA), stimulated with collagen types I, II, or III, or the control antigen streptokinase-streptodornase. Values for PGE_2 are expressed as a stimulation index (stimulated value/control unstimulated value). (Reprinted with permission from Dayer JM, Trentham DE, Krane SM: Collagens act as ligands to stimulate human monocytes to produce mononuclear cell factor (MCF) and prostaglandins (PGE_2). *Collagen Rel. Res.* 2:523–40, 1982.)

arachidonic acid metabolism by macrophagelike Langerhans cells of the skin, and specifically the synthesis of PGE_2 by these cells, is an important part of the healing process (53). There are indications that Langerhans cells participate in activation of suppressor lymphocytes *in situ* (54), probably by a PGE_2-mediated mechanism, which suggests a possible role of the lymphocyte in the healing process. Unfortunately, few data are currently available in this area. Studies are underway in several laboratories, however, to determine the local functions of phagocyte-produced PG.

LITERATURE CITED

1. Schneider E, Dy M: Activation of macrophages. *Comp. Immunol. Microbiol. Infect. Dis.* 8:135–46, 1985.
2. Fontana A, Bodner S, Frei K: Interleukin 1. *Schweiz Med. Wochenschr.* 115:1424–8, 1985.
3. Oppenheim JJ: *Interleukins and Interferons in Inflammation.* Current Concepts series, Upjohn Co., Kalamazoo, 1986.
4. Bonta IL, Parnham MJ: Immunomodulatory-antiinflammatory functions of E-type prostaglandins. Minireview with emphasis on macrophage mediated effects. *Int. J. Immunopharmacol.* 4:103–9, 1982.
5. Oropeza-Rendon RL, Bauer HC, Fischer H: Effect of prostaglandin E_1 on the level of cAMP in bone marrow macrophages. Inhibition of phagocytosis and cell shape changes. *J. Immunopharmacol.* 2:133–7, 1980.
6. Bonney RJ, Burger S, Davies P, Kuehl FA, Humes JL: Prostaglandin E_2 and prostacyclin elevate cyclic-AMP levels in elicited populations of mouse peritoneal macrophages. *Adv. Prostgl. Thrombox. Res.* 8:1691–3, 1980.
7. Bonta IL, Adolfs MJP, Parnham MJ: Prostaglandin E_2 elevation of cyclic AMP in granuloma macrophages at various stages of inflammation: relevance to antiinflammatory and immunomodulating functions. *Prostaglandins* 22:95–103, 1981.
8. Bonta IL, Adolphs MJP, Parnham MJ: Distinction between responsiveness of macrophages to cyclic-AMP elevation by prostaglandin E_2 and prostacyclin. *Scand. J. Rheumat.* (suppl.) 40:58–61, 1981.
9. Taffet S, Russell SW: Macrophage-mediated tumor cell killing: regulation of expression of cytolytic activity by prostaglandin E. *J. Immunol.* 126:424–9, 1981.
10. Schultz RM, Pavlidis NA, Stylos WA, Chirigos MA: Regulations of macrophage tumoricidal function: a role for prostaglandins of the E series. *Science* 202:320–7, 1978.
11. McCarthy ME, Zwilling BS: Differential effects of prostaglandins on the antitumor activity of normal and BCG-activated macrophages. *Cell. Immunol.* 60:91–7, 1981.
12. Schnyder J, Dewald B, Baggiolini M: Effects of cyclooxygenase inhibitors and prostaglandin E_2 macrophage activation in vitro. *Prostaglandins* 22:411–16, 1979.
13. Razin E, Bauminger S, Globerson A: Effect of prostaglandins on phagocytosis of sheep erythrocytes by mouse peritoneal macrophages. *J. Reticuloendothelial Soc.* 23:237–44, 1978.
14. Meerpohl HG, Bauknecht T: Role of prostaglandins on the regulation of macrophage proliferation and cytotoxic functions. *Prostaglandins* 31:961–71, 1986.

15. Snyder DS, Bellei DI, Unanue ER: Prostaglandins modulate macrophage Ia expression. *Nature* (London) 299:163–5, 1982.
16. Unanue ER: The regulatory role of macrophages in antigenic stimulation. II. symbiotic relationship between lymphocytes and macrophages. *Adv. Immunol.* 31:1–136, 1981.
17. Scher MG, Belber DI, Unanue ER: Demonstration of a soluble mediator that induces exudates rich in Ia-positive macrophages. *J. Exp. Med.* 152:1684–98, 1980.
18. Snyder DS, Lu CY, Unanue ER: Control of macrophage Ia expression in neonatal mice – role of a splenic suppressor cell. *J. Immunol.* 128:1458–65, 1982.
19. Ferraris VA, DeRubertis FR, Hudson TH, Wolfe L: Release of prostaglandin by mitogen- and antigen-stimulated leukocytes in culture. *J. Clin. Invest.* 54:378–86, 1974.
20. Kurland JI, Bockman R: Prostaglandin E production by human blood monocytes and mouse peritoneal macrophages. *J. Exp. Med.* 147:952–5, 1978.
21. Morley J, Bray MA, Jones RW, Nugteren DH, VanDorp DA: Prostaglandin and thromboxanes production by human and guinea pig macrophages and leukocytes. *Prostaglandins* 17:729–36, 1979.
22. Goldyne ME, Stobo JD: Synthesis of prostaglandins by subpopulations of human peripheral blood monocytes. Prostaglandins 18:687–95, 1979.
23. Kennedy MS, Stobo JD, Goldyne ME: In vitro synthesis of prostaglandins and related lipids by populations of human peripheral blood mononuclear cells. *Prostaglandins* 20:135–45, 1980.
24. Humes JL, Bonney RW, Pelus L, Dahlgren ME, Sadowski SS, Kuehl FA, Davies P: Macrophages synthesize and release prostaglandins in response to inflammatory stimuli. *Nature* (London) 269:149–51, 1977.
25. Brune K, Glatt M, Kalin H, Peskar B: Pharmacological control of prostaglandins and thromboxane release from macrophages. *Nature* (London) 274:261–3, 1978.
26. Bankhurst AD, Hastain E, Goodwin JS, Peake GT: The nature of the prostaglandin-producing mononuclear cell in human peripheral blood. *J. Lab. Clin. Med.* 97:179–86, 1981.
27. Foegh M, Maddox YT, Winchester J, Rakowski T, Schreiner G, Ramwell PW: Prostacyclin and thromboxane release from human peritoneal macrophages. *Adv. Prostagl. Thrombox. Leukotr. Res.* 12:45–9, 1983.
28. Gemsa D, Seitz M, Kramer W, Gruin W, Till G, Resch K: Ionophore A23187 raises cyclic AMP levels in macrophages by stimulating prostaglandin E formation. *Exp. Cell Res.* 118:55–62, 1979.
29. Resch K, Borillon D, Gemsa D: The activation of lymphocytes by the ionophore A23187. *J. Immunol.* 120:1514–20, 1978.
30. Greene WC, Parker CM, Parker CW: Calcium and lymphocyte activation. *Cell. Immunol.* 25:74–89, 1976.
31. Williams JD, Czop JK, Austen KF: Release of leukotrienes by human monocytes on stimulation of their phagocytic receptor for particulate activators. *J. Immunol.* 132:3034–40, 1984.

32. Neill MA, Henderson WR, Klebanoff SJ: Oxidative degradation of leukotriene C₄ by human monocytes and monocyte-derived macrophages. *J. Exp. Med.* 162:1634–44, 1985.

33. Schenkelaars EJ, Bonta IL: Cyclooxygenase inhibitors promote the leukotriene C₄ induced release of beta-glucuronidase from rat peritoneal macrophages: prostaglandin E suppresses. *Int. J. Immunopharmacol.* 8:305–11, 1986.

34. Feuerstein N, Foegh M, Ramwell PW: Leukotrienes C₄ and D₄ induce prostaglandin and thromboxane release from rat peritoneal macrophages. *Br. J. Pharmacol.* 72:389–91, 1981.

35. Goldyne ME, Stobo JD: T-cell-derived arachidonic acid and eicosanoid synthesis by macrophages. *Adv. Prostagl. Thrombox. Leukotr. Res.* 12:39–43, 1983.

36. Goodwin JS, Wiik A, Lewis M, Bankhurst AD, Williams RD: High affinity binding sites for prostaglandin E on human lymphocytes. *Cell. Immunol.* 43:150–9, 1979.

37. Goodwin JS, Kaszubowski PA, Williams RC: Cyclic adenosine monophosphate response to prostaglandin E₂ on subpopulations of human lymphocytes. *J. Exp. Med.* 150:1260–4, 1979.

38. Moretta L, Webb S, Grossi CE, Lydyard PM, Cooper MD: Functional analysis of two human T-cell subpopulations: help and suppression of B-cell responses by T cells bearing receptors for IgM or IgG. *J. Exp. Med.* 146:184–200, 1977.

39. Kaszubowski PA, Goodwin JS: Monocyte-produced prostaglandin induces Fcγ receptor expression on human T cells. *Cell. Immunol.* 68:343–8, 1982.

40. Spiegelberg HL: Biological activities of immunoglobulins of different classes and subclasses. *Adv. Immunol.* 19:259–94, 1974.

41. Melewicz FM, Zeiger RS, Mellon MH, O'Connor RD, Spiegelberg HL: Increased peripheral blood monocytes with Fc receptors for IgE in patients with severe allergic disorders. *J. Immunol.* 126:1592–95, 1981.

42. Fries LF, Mullins WW, Cho KR, Plotz PH, Frank MM: Monocyte receptors for the Fc portion of IgG are increased in systemic lupus erythematosus. *J. Immunol.* 132:695–700, 1984.

43. Ferreri NR, Howland WC, Spiegelberg HL: Release of leukotrienes C₄ and B₄ and prostaglandin E₂ from human monocytes stimulated with aggregated IgG, IgA, and IgE. *J. Immunol.* 136:4188–93, 1986.

44. Du JF, Foegh M, Maddox Y, Ramwell PW: Human peritoneal macrophages synthesize leukotrienes B₄ and C₄. *Biochim. Biophys. Acta.* 753:159–63, 1983.

45. Rouzer CA, Scott WA, Hamill AL, Liu FT, Katz DH, Cohn ZA: Secretion of leukotriene C and other arachidonic acid metabolites by macrophages challenged with immunoglobulin E immune complexes. *J. Exp. Med.* 156:1077–86, 1982.

46. Raukin JA, Hitchcock M, Merrill WW, Huang SS, Brashler JR, Bach MK, Askenase PW: IgE immune complexes induce immediate and prolonged release of leukotriene C₄ (LTC₄) from rat alveolar macrophages. *J. Immunol.* 132:1993–9, 1984.

47. Stringfellow DA, Brideau R: Interferons and prostaglandins. *Interferon* 2:147–63, 1984.
48. Schultz RM: E-type prostaglandins and interferons: yin-yang modulation of macrophage tumorcidal activity. *Med. Hypotheses* 6:831–43, 1980.
49. Boraschi D, Censini S, Bartalini M, Ghiara P, Simplicio PD, Tagliabue A: Interferons inhibit LTC_4 production in murine macrophages. *J. Immunol.* 138:4341–6, 1987.
50. Yoneda T, Mundy GR: Prostaglandins are necessary for osteoclast-activating actor production by activated peripheral blood leukocytes. *J. Exp. Med.* 149:279–83, 1979.
51. Wahl LM, Olsen CE, Wahl SM, McCarthy JB, Sandberg AL, Mergenhagen SE: Prostaglandin and cyclic AMP regulation of macrophage involvement in connective tissue destruction. *Ann. N.Y. Acad. Sci.* 332:271–8, 1979.
52. Dayer JM, Trentham DE, Krane SM: Collagens act as ligands to stimulate human monocytes to produce mononuclear cell factor (MCF) and prostaglandins (PGE_2). *Collagen Rel. Res.* 2:523–40, 1982.
53. Penneys N: *Prostaglandins and the Skin.* Current Concepts series, Upjohn Co., Kalamazoo, 1980.
54. Granstein RD, Askari M, Whitaker D, Murphy GF: Epidermal cells in activation of suppressor lymphocytes: further characterization. *J. Immunol.* 138:4055–62, 1987.

4

Lymphocyte response

It is now clear that the immunological activities of lymphocytes are profoundly affected by the presence of various products of arachidonic acid metabolism; however, many fundamental issues concerning this interaction remain unresolved and are, at times, hotly debated.

EICOSANOID PRODUCTION AND INTERACTION WITH LYMPHOCYTES

It has not been proven conclusively that normal lymphocytes possess either cyclooxygenase or lipoxygenase enzyme systems. This is stated with caution, since studies *in vitro* by Parker et al. (1) suggested that purified lymphocytes produced 5-HETE, 12-HETE, and TXB$_2$, and Goetzl reported that T lymphocytes possess a 5-lipoxygenase system (2). Furthermore, Bauminger reported that mouse thymocytes could be differentiated on the basis of their PG synthetase activities (3). Mature thymocytes appeared to contain higher concentrations of PGE and greater PG synthetase activity than did immature cells. Webb and Nowowiejski reported that both nonadherent and glass-adherent T cells release PGE into the culture medium after phytohemagglutinin (PHA) stimulation (4).

Using thin layer chromatography (TLC) and/or radioimmunoassay (RIA) procedures, Aussel et al. recently demonstrated that Jurkat cells (a human leukemic T cell line) are able to convert arachidonic acid to 6-keto-PGF$_{1\alpha}$, PGE$_2$, PGA$_2$, and TXB$_2$ (5). The addition of lectins or anti-CD3 monoclonal antibody reduced the amount of PG released by the cells, as did the addition of the cyclooxygenase inhibitors indomethacin and niflumic acid. These results are shown in Figure 4.1. Aussel et al. suggested that endogenous PG synthesis by lymphocytes might serve to keep them in a quiescent state, while reduction in PG synthesis is required as part of the activation sequence (5).

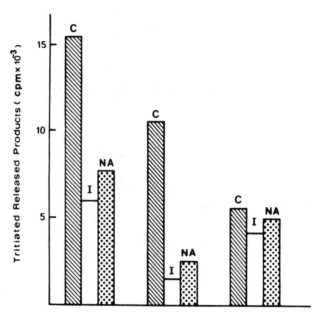

Figure 4.1. PG synthesis by Jurkat cells in the absence (control = c) or presence of 10μm indomethacin (IM) or 10μm niflumic acid (NA). Cells were labeled for 24 h with [³H] AA, washed, then incubated for an additional 4 h. Each bar represents the mean of two experiments, performed in triplicate. The left-hand set of three bars represents total tritiated products released: the middle set represents PG synthesis; the right hand set represents arachidonic acid not utilized. (Reprinted with permission from Aussel C, Didier M, Fehlmann M: Prostaglandin synthesis in human T cells: its partial inhibition by lectins and antiCD₃ antibodies as a possible step in T cell activation. *J. Immunol.* 138:3094–9, 1987.)

Other results, however, suggest that normal lymphocytes do not produce PGs or leukotrienes (LTs). Studies by Kennedy et al. (6), Bankhurst et al. (7), Goldyne et al. (8), and others, suggest that the presence of eicosanoids in lymphocyte culture supernatants is often caused by the presence of contaminating monocytes and/or platelets. But, as Goodwin and Webb have pointed out, the techniques most often used to eliminate monocytes all depend on cell adherence, making it impossible at this time to rule out the existence of a PG-producing, glass-adherent T cell (9). Clearly, continued studies are needed in this area.

Further confusion has been created in the literature through the interpretation of results of experiments using a multitude of nonsteroidal antiinflammatory agents, some of which are much less specific inhibitors of PG synthesis than was once thought. A good example is indomethacin. It is now known that even submicromolar concentrations of this drug do not specifically inhibit PG synthetase activity. Instead, 10^{-8} M concentrations also inhibit cyclic-AMP-dependent protein kinase activity (10) and higher concentrations also inhibit phosphodiesterase (11). This clarification of indomethacin activity makes necessary the reevaluation of a great quantity of literature produced over the last 5–10 years.

Another fundamental (and unresolved) issue is the actual importance of the interaction between lymphocytes and the products of arachidonic acid metabolism. The role of arachidonic acid metabolites in homeostasis is clear in the case of inflammation (12). Although still debated, there are now strong indications that these same metabolites are involved in a similar homeostasis of the specific immune response because:

1. prostaglandin production is now known to be an integral part of cell activation;
2. an interrelationship between PGs and LTs, and other immunoregulatory monokines and lymphokines is clearly indicated by recent research; and
3. there is irrefutable evidence that arachidonic acid metabolites are potent activators of regulatory lymphocyte subpopulations, though dose and timing are critical variables.

The recent introduction of a variety of specific inhibitors of enzymes of the arachidonic acid cascade, as well as the commercial availability of many of the metabolites themselves, have allowed more definitive studies to be conducted than were previously possible, thereby giving a better appreciation for the profound effects that the PGs and LTs can have on lymphocyte responsiveness. Even so, it is still argued that though PGs may be important in various disease states, their role in normal individuals must be relatively minor as the immune response of humans and animals given long-term and sometimes high-dose cyclooxygenase inhibitors appears to be quite normal (13). Recent studies have shown, however, that at less than toxic doses, nonsteroidal antiinflammatory agents can only inhibit approximately 40–60% of cyclooxygenase activity *in vivo* (Tables 4.1 and 4.2). While results vary somewhat from tissue to tissue, these results nevertheless suggest that even in the face of long-term medication, sufficient arachidonate metabolism remains to fulfill normal immunological roles (Figure 4.2) (14,15).

Prostaglandin-receptor–adenylate-cyclase interaction

In order for a PG to exert a biological effect (a) an appropriate receptor-response circuit must exist in the responding cells, and (b) the PG must be

Table 4.1. *Systemic venous plasma PG levels in sheep, measured before and after administration of indomethacin (2–10 mg/kg).*

Prostaglandin	n	Control	Time after onset of indomethacin infusion		
			5 min	20 min	180 min
E_2 (pg/ml)	14	214 ± 18	200 ± 26	135 ± 19	112 ± 10
%Δ			−6 ± 8	−36 ± 6	−45 ± 5
F_{2a} (pg/ml)	10	180 ± 17	111 ± 9	87 ± 19	65 ± 10
%Δ			−37 ± 3	−50 ± 9	−57 ± 12

Source: Reprinted with permission from Naden RP, Iliya CA, Arant BS, Grant NF, Rosenfeld CR: Hemodynamic effects of indomethacin in chronically instrumented pregnant sheep. *Am. J. Obstet. Gynecol.* 151:484–94, 1985.

Table 4.2. *Plasma PG levels in the vena cava and uterine vein in sheep, before and after indomethacin administration (2–10 mg/kg).*

Prostaglandin	Control	Time after onset of indomethacin infusion		
		5 min	20 min	180 min
E_2 (pg/ml)				
Vena cava	233 ± 30	222 ± 48	181 ± 47	113 ± 18
%Δ		−5 ± 14	−27 ± 11	−42 ± 11
Uterine vein	458 ± 67	531 ± 109	225 ± 28	197 ± 22
%Δ		+13 ± 6	−50 ± 5	−54 ± 7
F_{2a} (pg/ml)				
Vena cava	188 ± 4	112 ± 5	126 ± 29	66 ± 14
%Δ		−40 ± 3	−41 ± 11	−64 ± 7
Uterine vein	223 ± 22	191 ± 22	90 ± 23	105 ± 11
%Δ		−13 ± 8	−58 ± 12	−52 ± 5

Source: Reprinted with permission from Naden RP, Iliya CA, Arant BS, Grant NF, Rosenfeld CR: Hemodynamic effects of indomethacin in chronically instrumented pregnant sheep. *Am. J. Obstet. Gynecol.* 151:484–94, 1985.

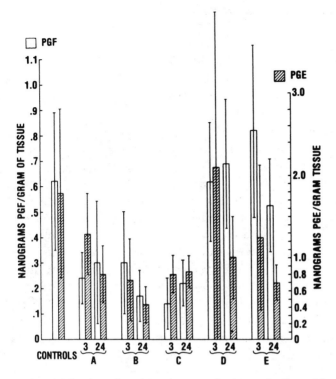

Figure 4.2. PG levels as found in one example tissue, the liver, 3 h and 24 h following oral administration of nonsteroidal antiinflammatory agents (A) ibuprofen, (B) flurbiprofen, (C) indomethacin, (D) 3 acetonitrile,4,5-bis(*p*-methoxyphenyl)-2-phenyl-pyrrole, and (E) acetaminophen. (Reprinted with permission from Fitzpatrick FA, Wynalda MA: In vivo suppression of prostaglandin biosynthesis by nonsteroidal antiinflammatory agents. *Prostaglandins.* 12:1037–51, 1976.)

present in a high enough concentration to occupy a sufficient number of receptors to trigger a response. It has been reported that the concentration of biologically active PGs in the normal circulation is approximately 10^{-10} M (16,17). The minimum PG concentration necessary to elicit most biological responses, however, is 10^{-8}–10^{-9} M (18,19). It seems logical, therefore, that the eicosanoids exert their greatest effect at the site of synthesis, and before dilution in the circulation.

It has been proposed that most, if not all, of the biological effects of the PGs and thromboxanes are dependent upon changes in membrane 3', 5' cyclic adenosine monophosphate levels. The structure and biological activity of the LTs suggest that they may also influence cyclic nucleotide levels, and this has been found to be the case in recent studies by Mexmain et al. (20).

Figure 4.3 is a simplified illustration of the interaction of a PG with its cell surface receptor. The lymphocyte membrane has been demonstrated to bear varying numbers of receptors for a number of arachidonic acid metabolites including PGE (21). The coupling of the receptor with its PG ligand stimulates adenylate cyclase activity, which catalyzes the conversion of ATP into cyclic AMP. Cyclic AMP then binds to the regulatory subunit of a protein kinase. The catalytic subunit of the protein kinase is then activated, which leads to subsequent phosphorylation of a variety of proteins. The phosphorylation process in turn can stimulate or inhibit the expression of a specific biological effect by protein products.

There is ample evidence for the existence of specific receptors for PGE, PGA, PGF, and PGI on various cells, as well as indirect evidence for a TXA_2 receptor on platelets and blood vessels. Table 4.3 lists some of the cells and tissues having binding sites for PGE, a PG of particular immunological interest, as well as dissociation constants for the prostaglandin-receptor interaction and subsequent cyclic AMP levels. PGE receptors generally have dissociation constants in the $10^{-8}-10^{-9}$ range and Scatchard analysis occasionally results in a biphasic or curvilinear plot that indicates heterogeneity of binding sites (22).

Rao reviewed the relationship between prostaglandin-receptor binding and the expression of biological activity (23) and concluded that, as is the case with many hormones, receptor binding and response were not well correlated. Using cultured cells (a murine fibroblast clone), however, Brunton et al. found that PGE_1 binds with an affinity of 0.5×10^{-10} M, an affinity which correlated well with concentration-dependent activation of adenylate cyclase (24).

Goodwin et al. identified high-affinity binding sites for PGE on human lymphocytes, which showed specific and reversible binding of [^{3}H] PGE_1 and PGE_2 but not other PGs (21). They calculated a dissociation constant (K_d) of 2×10^{-9} M and a Bmax of approximately 200 binding sites per cell. Goodwin et al. also found that large quantities of PGA, $PGF_{1\alpha}$ or $PGF_{2\alpha}$ did not inhibit the binding of [^{3}H]PGE (21).

As for the activation of adenylate cyclase, however, there is a discrepancy between the PGE concentrations that have been reported to increase cyclic AMP levels ($10^{-7}-10^{-4}$ M) (25,26) and the PGE concentrations that produce changes in lymphocyte function ($10^{-10}-10^{-8}$ M) (13,27), suggesting that other metabolic pathways may be involved in PGE-lymphocyte interaction. It is relevant to note that cyclic AMP is not the

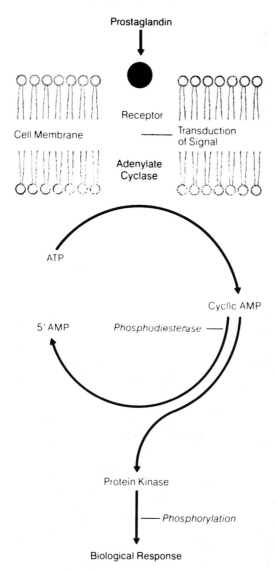

Figure 4.3. Simplified schematic drawing of PG interaction with its cell surface receptor, and the biological consequences of that interaction. (Reprinted with permission from Gorman RR, Marcus AJ: *Prostaglandins and Cardiovascular Disease.* Current Concepts. Upjohn, Kalamazoo, p. 10, 1981.)

Table 4.3. *A review of some of the defined PG binding sites, their dissociation constants, and their relationship to cyclic AMP.*

Species & Tissue	Preparation	Apparent Kd (M)	PGE, or PGI₂-Induced Change in Cyclic AMP
Rat Adipocyte	Homogenate	3.0×10^{-9}	↓
Rat Adipocyte	Plasma Membrane "Ghosts"	2.0×10^{-10} & 2.5×10^{-9}	↑
Rat Liver	Plasma Membrane	1.0×10^{-9} & 2.5×10^{-8}	↑
Human Platelets	Plasma Membrane & Intact Cells	8.0×10^{-11} & 1.2×10^{-8}	↑
Cultured Murine L Cells	Homogenate	2.0×10^{-8}	↑
Rat Thymocyte	Plasma Membrane	7.0×10^{-11} & 2.0×10^{-9}	Data not available
Rat Skin	Plasma Membrane	1.5×10^{-7} & 1.9×10^{-7}	↑
Bovine Corpus Luteum	Plasma Membrane	1.3×10^{-9} & 1.0×10^{-8}	↑
Hamster Uterus	Homogenate	7.7×10^{-10} & 1.3×10^{-8}	↑
Human and Ovine Adrenal	Homogenate	1.5×10^{-8}	↑

Source: Reprinted with permission from Gorman RR, Marcus AJ: Prostaglandins and cardiovascular disease. Current Concepts Series. Upjohn Co., Kalamazoo, p. 11, 1981.

second messenger for PGE in every tissue. For example, studies by Sinha and Colman demonstrated that PGE_1 inhibits platelet aggregation by a cyclic-AMP-independent pathway (28) and Bito described an active uptake system for PGE in many tissues, which is distinct from membrane receptors coupled to adenylate cyclase and yet is involved in mediating many PG effects (29).

The best work in this area, however, supports the hypothesis that cyclic AMP is the second messenger for PGE_2 in the human lymphocyte because:

1. the concentrations of PGE_2 that increased cyclic AMP were the same as those concentrations suppressing mitogenesis and concentrations at which [^{3}H]PGE_2 binds to lymphocytes;
2. the order of the potency of the various prostaglandins with respect to the cyclic AMP response paralleled the order of potency of these compounds as inhibitors of lymphocyte mitogenesis; and
3. PGE_2 was shown to stimulate cyclic AMP and inhibit blastogenesis of fresh lymphocytes, while preincubated lymphocytes were neither suppressed nor did they display an increase in cyclic AMP (Figure 4.4) (30).

ARACHIDONIC ACID METABOLISM AND T-CELL FUNCTION

As early as 1971, Smith et al. reported that several prostaglandins were capable of inhibiting PHA stimulation of human T lymphocytes, as measured *in vitro* by [^{3}H]thymidine incorporation (31). Within a few years, other investigators had reported that other assay systems *in vitro* showed the inhibitory properties of PGE_1 and PGE_2, including depression of macrophage inhibitory factor activity (32), leukocyte inhibitory factor production (33), cytolysis of activated lymphocytes (25), inhibition of hemolytic plaque formation (34), and reduced antibody production (35). Many of these studies were flawed, however, by the high concentrations (10^{-6}–10^{-4} M) of PGs that were used.

PGE_2 concentrations are considered to be in the physiological range of 10^{-8} M or less, as these are the concentrations found at sites of inflammation (12). More recent studies have shown that those physiological concentrations of PGE_2 will suppress most of the manifestations of T-cell function including:

mitogen responsiveness (36,37)
clonal proliferation (38,39)
antigenic stimulation (40)
E-rosette formation (41,42)
lymphokine production (43-46)
generation of cytotoxic cells (47,48)
lymphocyte migration (49)

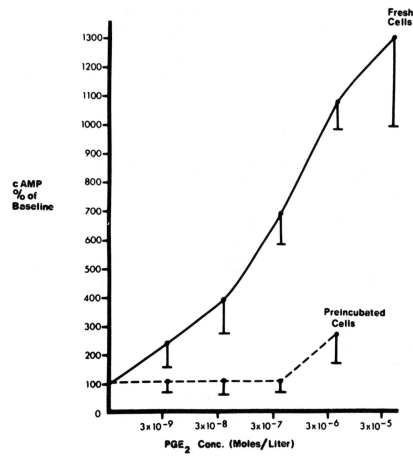

Figure 4.4. The effect of 20 h PGE_2 preincubation on the cyclic AMP response of peripheral blood mononuclear cells. Data are shown as mean ± SEM for five subjects, each measured in duplicate. The cyclic AMP levels of preincubated cells are significantly elevated only after exposure to 3×10^{-6} M PGE_2 ($p < 0.01$). (Reprinted with permission from Goodwin JS, Bromberg S, Messner RP: Studies on the cylic AMP response to prostaglandin in human lymphocytes. *Cell. Immunol.* 60:298–307, 1981.)

Eicosanoids and in vitro mitogen response

The assay probably used most often to study prostaglandin–T-lymphocyte interactions has been the blastogenic response *in vitro* of T cells induced by PHA, or concanavalin A (con A). PHA-stimulated cultures of

Table 4.4. *The increase in [³H]thymidine incorporation caused by indomethacin at different PHA concentrations.*

PHA concentration	Before indomethacin	Increase with indomethacin
µg/ml	*cpm*	*%*
– 0 –	142 ± 47	70 ± 48 (*P* < 0.001)
2	4,078 ± 5,852	55 ± 46 (*P* < 0.001)
10	12,630 ± 6,448	58 ± 28 (*P* < 0.001)
20	13,334 ± 7,573	46 ± 28 (*P* < 0.001)
50	7,984 ± 7,657	16 ± 16 (*P* < 0.01)

Note: Each data point represents the mean ± SD of 15 subjects. Cells were cultured with or without 1.0 µg/mL indomethacin as indicated.

Source: Reprinted with permission from Goodwin JS, Bankhurst AD, Messner RP: Suppression of human T-cell mitogenesis by prostaglandin: existence of a prostaglandin-producing suppressor cell. *J. Exp. Med.* 146:1719–34, 1977.

human peripheral blood mononuclear cells have been found to produce endogenous PGE_2 at concentrations that are suppressive when added from exogenous sources (9). Goodwin et al. showed that such concentrations (10^{-9}–10^{-6} M) of PGE_1 and PGE_2 – but not PGA, $PGF_{1\alpha}$, or $PGF_{2\alpha}$ – were able to significantly inhibit PHA- and con-A-induced [³H]thymidine incorporation (50). Because of the endogenous production of PGE by mitogen-stimulated mononuclear cell cultures, the addition to cultures of PG synthetase inhibitors, such as indomethacin, generally results in increased cell proliferation, as shown in Table 4.4. There is general agreement that the monocyte/macrophage is the source of endogenous PG in these cultures, and it has been noted that lymphocyte cultures are more sensitive to PGE when mitogen concentrations are lower (51).

Similarly, T-lymphocyte proliferation in mixed lymphocyte cultures is also suppressed by endogenous and exogenous prostaglandins (47,52). Table 4.5 summarizes the effect of adding 10, 100, or 1000 ng/ml of individual synthetic PGs to normal, human mixed lymphocyte cultures. Statistically significant suppression of the T-cell response (p <0.01) at physiological concentrations (10^{-8} M) was seen only with PGE_2. At higher concentrations (10^{-6} M) significant suppression occurred also with the addition of PGE_1 and PGA_1. Other PGs appeared to have no significant effect on T lymphocytes in this assay system (52).

Recent studies have suggested that several products of lipoxygenase activity also control cellular immune responses. Goodman and Weigle

Table 4.5. *The effect of synthetic PG and thromboxane additions upon normal human mixed lymphocyte responsiveness.*

Added to Cultures	Quantity	DPM ± SD	% sp/st
Control	—	36,375 ± 2,607	—
PGA$_1$	10 ng/ml	37,740 ± 4,819	4 st
	100 ng/ml	34,652 ± 11,730	5 sp
	1,000 ng/ml	7,607 ± 261	79 sp
PGA$_2$	10 ng/ml	29,017 ± 1,100	20 sp
	100 ng/ml	28,263 ± 5,980	22 sp
	1,000 ng/ml	15,455 ± 1,423	42 sp
PGE$_1$	10 ng/ml	32,948 ± 5,937	9 sp
	100 ng/ml	28,551 ± 6,582	21 sp
	1,000 ng/ml	18,797 ± 5,712	48 sp
PGE$_2$	10 ng/ml	22,390 ± 676	38 sp
	100 ng/ml	21,278 ± 2,761	41 sp
	1,000 ng/ml	9,523 ± 2,619	77 sp
PGF$_1$ alpha	10 ng/ml	34,963 ± 2,134	4 sp
	100 ng/ml	35,158 ± 10,689	3 sp
	1,000 ng/ml	34,320 ± 5,888	6 sp
6-keto PGF$_1$ alpha	10 ng/ml	31,647 ± 4,548	13 sp
	100 ng/ml	32,898 ± 7,977	10 sp
	1,000 ng/ml	31,329 ± 7,095	14 sp
PGF$_2$ alpha thrometh.	10 ng/ml	31,479 ± 4,660	13 sp
	100 ng/ml	40,497 ± 4,301	11 st
	1,000 ng/ml	41,371 ± 3,817	14 st
Thromboxane B$_2$	10 ng/ml	33,303 ± 5,267	8 sp
	100 ng/ml	34,485 ± 5,350	5 sp
	1,000 ng/ml	31,326 ± 8,079	14 sp

Note: Statistically significant suppression of the response was observed only at the highest tested concentrations of PGA$_1$, PGA$_2$, and PGE$_1$, and at all three concentrations of PGE$_2$ ($p<0.01$). % sp/st, percent suppression or stimulation of the MLR; DPM ± SD, disintegrations per minute ± SD.

Source: Reprinted with permission from Ninnemann JL, Stockland AE: Participation of prostaglandin E in immunosuppression following thermal injury. *J. Trauma* 24:201–7, 1984.

reported that 15-HPETE inhibited mitogen-induced blastogenesis of mouse splenocytes (53) and Gualde et al. showed that this same metabolite inhibited E-rosette formation and mitogen-induced [^{3}H]thymidine incorporation by human T cells (54). Payan and Goetz (55) and Rola-Pleszczynski et al. (56) found that LTB$_4$ inhibited proliferation and lymphokine production *in vitro* by human lymphocytes. While Payan and Goetzl reported that LTC_4 and LTD_4 had no effect on human lymphocytes, Webb et al. found that as little as 10^{-12} M LTD_4 or LTE_4 significantly inhibited [^{3}H]thymidine incorporation by mitogen-stimulated mouse splenocytes (57). The suppressive activity of these eicosanoids appeared to be mediated through the stimulation of OKT8(+) suppressor T lymphocytes (58).

Lymphocyte activation by PGE_2

In addition to the now well-documented inhibitory activity of various eicosanoids, it appears that under certain conditions, PGE_2 is capable of activating lymphocytes. Stobo et al. showed that PGE_2 increases the mitogen-induced blastogenesis of low-density T lymphocytes, and decreases the response of medium-and high-density T cells (59). Gualde and Goodwin demonstrated that mitogen-induced proliferation of OKT8(+) T cells was increased after preincubation with PGE (60). Exposing peripheral blood mononuclear cells to PGE_2, followed by preincubation, results in increased NK activity, while PGE_2 added during NK–target-cell interaction is inhibitory (61).

Further indications that PGE may under certain circumstances participate in the activation of T lymphocytes arose in studies by Mertin and Stackpoole, who found that anti-PGE antisera significantly suppressed experimental allergic encephalomyelitis (EAE) in rats, and prevented the generation of host-versus-graft (HVG) and graft-versus-host (GVH) reactions in mice (62). They concluded that PGE may be important in the induction of cell mediated immune responses both *in vitro* and *in vivo*.

PGE_2 has been reported to induce immature thymocytes to differentiate into mature T cells (63). Thymic hormones may influence immature cells through stimulation of PGE synthesis (64) and, under at least some conditions, thymic hormone activity can be blocked by means of nonsteroidal antiinflammatory agents (65). Goodwin and Ceuppens have, therefore, concluded that PGE concentration, the activation and differentiation state of the target cell, and such variables as the length of exposure of the target cell to PGE, critically influence the outcome of the interaction, which can vary from suppression to stimulation (9).

Induction of suppressor cell activity

Various products of arachidonic acid metabolism participate in the induction of nonspecific suppressor T lymphocyte activity. There is ample evidence that PGE_2 is particularly important in this regard. Fisher et al. studied the immunoregulatory activity of human macrophages on B- and T-cell proliferation and found that monocyte suppression was predominantly mediated through the elaboration of PGE_2, and was dependent upon the presence of a "short-lived" radiosensitive T-lymphocyte subpopulation (65). Furthermore, Fisher et al. showed that high concentrations ($>10^{-7}$ M) PGE_2 alone can activate T-suppressor cells, which display the same characteristics as con-A-activated T-suppressor cells. The addition of antiPGE$_2$ antiserum or indomethacin prevented the activation of con A suppressors, showing that a double signal was necessary for con A to have an effect on suppressor cell development. Presumably con A renders suppressor precursor cells to become sensitive to small quantities of PGE_2 produced by monocytes, leading to suppressor induction (66).

Kaszubowski and Goodwin showed that monocyte-produced PGE_2 (3×10^{-6} M) induced Fcγ receptor expression on human T cells, and that these cells were responsible for suppressor activity *in vitro* (67). Supportive results have been reported by Rappaport and Dodge (68) and by Chouaib et al. (69). The latter group found that, when cell separation was performed before incubation with PGE_2, OKT8(+) lymphocytes were responsible for 58% of the observed suppression; 27% of the suppression was associated with OKT4(+) lymphocytes. When cell separation was carried out after PGE_2 incubation, however, OKT8(+) lymphocytes were responsible for an even greater portion of the suppressive activity (68%) whereas OKT4(+) lymphocytes had a reduced capacity to suppress (15%). While PGE_2 treatment of peripheral blood lymphocytes induced suppressor T cell activity, PGE_2 incubation was not found to alter the distribution of T-cell subsets (Table 4.6) (69).

Fulton and Levy likewise found that PGE (in this case PGE_1) was capable of inducing nonspecific suppressor cell activity in murine splenocytes (70). Induction of this suppressor cell required the presence of Thy-1 bearing cells, but did not require nylon-wool adherent cells for activation. Results obtained following the addition of indomethacin to cultures suggest that, while PG was necessary for the induction of suppressor activity, no further PG synthesis was necessary for the expression of that activity (70).

Webb and Nowowiejski showed induction of suppressor cell activity by the addition of PGE_2 to murine spleen cells, and identified the cells as glass-adherent T lymphocytes (4). PGE_2 induced the production of a

Table 4.6. *Incubation of enriched T lymphocytes for 48 h in the presence of PGE_2 does not modify their OKT4 and OKT8 antigenic markers.*

Cells Tested	Monoclonal Antibodies	% Fluorescence with Monoclonal Antibody		
		Expt. 1	Expt. 2	Expt. 3
48 hr PGE_2-treated T cells	OKT3	96	95	96
	OKT4	54	60	53
	OKT8	30	34	32
48 hr medium-treated T cells	OKT3	97	95	95
	OKT4	58	62	55
	OKT8	27	30	27

Note: Enriched T cells were preincubated in the presence of PGE_2 (10^{-6} M) for 48 h at 37 °C, or in control medium. The cells were washed and binding of monoclonal antihuman T cell antibody was determined by immunofluorescence microscopy.
Source: Reprinted with permission from Chouaib S, Chatenoud L, Klatzmann D, Fradelizi D: The mechanisms of inhibition of human IL-2 production. II. PGE_2 induction of suppressor T lymphocytes. *J. Immunol.* 132:1851–7, 1984.

"suppressor product" by these cells, which characterized, not as PGE, but rather as a heat-stable protein (4).

In addition to PGE_2-induction of suppressor T cells, there is evidence that 5- and 15-lipoxygenase products also participate in this important function. Preincubation of T cells with LTB_4 in nanomolar or picomolar concentrations confers on them an ability to suppress immunoglobulin (Ig) production in PWM-stimulated cultures of fresh mononuclear cells (71). The suppressor cell induced by LTB_4 is OKT8(+), radiosensitive, and its generation can be blocked by cycloheximide. When OKT8(−) T lymphocytes were incubated with LTB_4 for 18 h, 10–20% of the cells became OKT8(+), a process blocked by cycloheximide but not by mitomycin C (71). Most striking in this effect is the quantity of LTB_4 necessary to induce T-suppressor activity. Gualde and Goodwin report that as little as 10^{-12} M LTB_4 causes significant inhibition of IgG production when added directly to PWM-stimulated cultures or when used to activate T-suppressor cells by preincubation. This indicates that LTB_4 is three to six orders of magnitude more potent as an inducer of suppressor activity than either PGE or histamine (72). These results are summarized in Table 4.7.

Murine lymphocytes preincubated with either 15-HPETE or 15-HETE partially inhibited T-cell proliferation (73). Lymphocytes incubated *in vitro* for 48 h with 10^{-5}–10^{-8} M 15-HETE had increased expression of

Table 4.7. *The effect of preincubating T cells for 18 h with LTs on the production of IgG and IgM in subsequent PWM-stimulated cultures of fresh autologous B cells.*

Cells	T cells preincubated with	IgG (ng/ml)	Percent inhibition	IgM (ng/ml)	Percent inhibition
B	...	93 ± 10	...	23 ± 6	...
B + T	0	1101 ± 124	...	315 ± 39	...
B + T	10^{-12} M LTB$_4$	901 ± 93	18 ± 4	285 ± 31	9 ± 3
B + T	10^{-11} M LTB$_4$	752 ± 112	32 ± 4	208 ± 27	34 ± 4
B + T	10^{-10} M LTB$_4$	531 ± 102	52 ± 5	214 ± 14	32 ± 3
B + T	10^{-9} M LTB$_4$	348 ± 105	68 ± 5	191 ± 19	39 ± 3
B + T	10^{-8} M LTB$_4$	258 ± 98	76 ± 4	129 ± 17	59 ± 4
B + T	10^{-9} M LTC$_4$	963 ± 84	12 ± 3	287 ± 46	9 ± 3
B + T	10^{-8} M LTC$_4$	957 ± 96	13 ± 6	288 ± 33	8 ± 4
B + T	10^{-9} M LTD$_4$	991 ± 114	10 ± 4	291 ± 34	8 ± 4
B + T	10^{-8} M LTD$_4$	909 ± 106	17 ± 4	278 ± 41	12 ± 3

Source: Reprinted with permission from Gualde N, Goodwin JS: Induction of suppressor T cells in vitro by arachidonic acid metabolites issued from the lipoxygenase pathway. *In*: Bailey JM (ed.), *Prostaglandins, Leukotrienes, and Lipoxins.* Plenum Press, N.Y., pp 565–75, 1985.

Lyt-2 antigen (74). Similarly, human T cells preincubated with 15-HPETE suppressed Ig production by fresh autologous PWM-stimulated B cells (75). When 15-HPETE-preincubated human T lymphocytes were treated with anti-OKT8 monoclonal antibody plus complement, suppressive effects were eliminated. Removal of the OKT8(+) cells prior to preincubation with 15-HPETE, however, did not prevent the generation of new OKT8(+)-expressing T lymphocytes, presumably from OKT8(−) presuppressor T cells (75). This interpretation is supported by experiments by Gualde et al. with isolated OKT8(−) T-cell populations, which could be converted to OKT8(+) cells via exposure to 15-HPETE (75).

Regulation of lymphokine production and response

Chouaib and Fradelizi showed that IL-2 production by PHA-stimulated human T lymphocytes could be inhibited by the addition of monocytes to the cultures (76). IL-2 inhibition was mediated by soluble monocyte products, including PGE$_2$. When T lymphocytes were irradiated, IL-2 production was no longer affected by the addition of monocytes or their culture supernatants. To explain these results, Chouaib and Fradelizi proposed that PGE$_2$ activates a subset of T cells, which is capable of

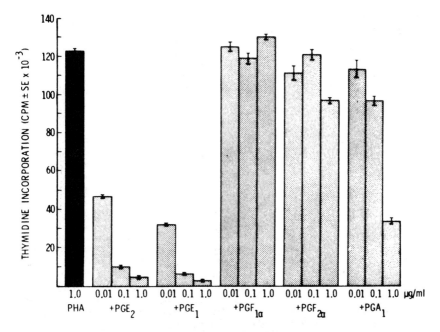

Figure 4.5. The effect of various PGs on PHA-induced production of IL-2 by lymphocytes from a representative normal human donor. (Reprinted with permission from Rappaport RS, Dodge GR: Prostaglandin E inhibits the production of human interleukin 2. *J. Exp. Med.* 155:943–8, 1982.)

suppressing IL-2 production (76). Later work proved that this was indeed the case, thereby explaining at least one of the modes of action of PGE_2-induced suppressor cells (69,77).

Rappaport and Dodge confirmed that exogenous PGE was capable of inhibiting IL-2 production by normal human peripheral blood lymphocytes, while PGF and PGA did not have this effect (Figure 4.5) (45). Removal of glass-adherent cells from mononuclear cell populations led to increased IL-2 production by nonadherent cells (Table 4.8), and the addition of PG synthetase inhibitors raised IL-2 production to supranormal levels (45). PGE also appears to inhibit lymphokine production by guinea pig lymph node lymphocytes (78) and IL-2 production by murine lymphocytes (79).

In addition to regulating the production of IL-2 by T lymphocytes, there is now ample evidence that PGE_2 can suppress the expression of IL-2 receptors, and the response of lmphocytes to the addition of exogenous IL-2. This has been demonstrated in a mouse model (79) and

Table 4.8. *The effect of the removal of adherent cells from mononuclear cell populations on the production of IL-2.*

Donor	Interleukin 2-mediated lymphocyte transformation [³H]thymidine incorporation[a]	
	Unfractionated mononuclear cells[b]	Nonadherent mononuclear cells[c]
	cpm ± SE	
A	34,062 ± 1,159	87,711 ± 2,433
B	54,870 ± 4,497	78,880 ± 3,475
C	14,092 ± 541	135,336 ± 2,825
D	8,080 ± 164	61,187 ± 2,667
E	6,367 ± 426	66,587 ± 3,433
F	6,781 ± 418	166,093 ± 7,084
G	3,344 ± 234	88,400 ± 2,759
H	30,085 ± 865	155,308 ± 2,390

[a] 18–20 h before harvest the cells were labeled with 1 μCi/ml [³H]thymidine (>15 Ci/mmol). The results represent the mean of quadruplicate determinations for each condition. The mean response of assay cells to medium and to medium containing fresh PHA (1 μg/ml) was 1,556 ± 356 cpm and 3,646 ± 949 cpm, respectively, over all experiments. The response of assay cells to conditioned medium from donor-paired unfractionated and nonadherent cell populations was found to be significantly different by two-way analysis of variance ($P < 0.001$) for all donors.
[b] The mean response of assay cells to supernatants from resting unfractionated lymphocytes was 1,864 ± 244 cpm ($n = 8$).
[c] The mean response of assay cells to supernatants from resting nonadherent lymphocytes was 1,797 ± 348 cpm ($n = 8$).

Source: Reprinted with permission from Rappaport RS, Dodge GR: Prostaglandin E inhibits the production of human interleukin 2. *J. Exp. Med.* 155:943–8, 1982.

is of particular clinical relevance in the immunological changes that take place in patients with major systemic injuries (80,81).

Prostaglandins also affect the production of IL-1 by monocytes and macrophages, an event fundamental to the initiation and maintenance of immune responses. Kunkel et al. showed, in a mouse model, that the addition of exogenous PGE_2 and prostacyclin (PGI_2), but not $PGF_{2\alpha}$, suppressed IL-1 production in a dose-dependent manner by LPS-stimulated macrophages (82). Conversely, the addition of cyclooxygenase inhibitors produced dose-dependent augmentation of IL-1 production (Figure 4.6) (82). Inhibitors of lipoxygenase activity appeared to have little effect on LPS-induced IL-1 production. The clinical relevance of these observations can be illustrated by recent work by Korn et al. with a 30 000 dalton IL-1 inhibitor derived from the urine of febrile patients (83). This material was found to stimulate PGE_2 synthesis by fibroblasts (Figure 4.7), as well as to block IL-1-mediated stimulation of thymocyte

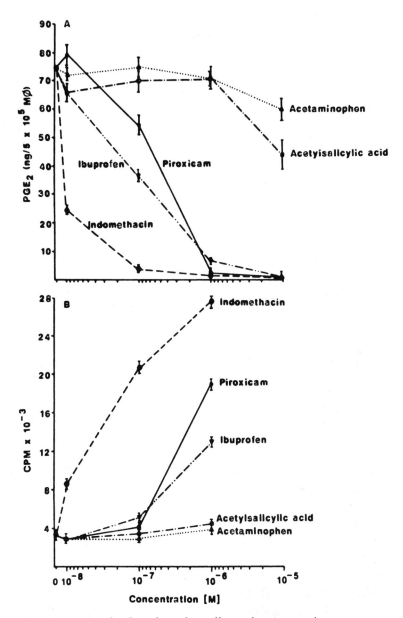

Figure 4.6. (A) The dose-dependent effects of various cyclooxygenase inhibitors on the synthesis of PGE_2 and (B) the production of Il-1 by LPS-stimulated murine resident peritoneal macrophages. (Reprinted with permission from Kunkel SL, Chensue SW, Phan SH: Prostaglandins as endogenous mediators of interleukin 1 production. *J. Immunol.* 136:186–92, 1986.)

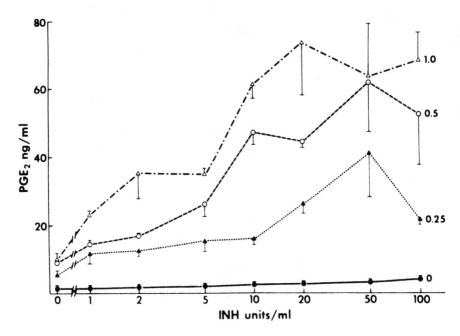

Figure 4.7. Dose–response curve of IL-1 inhibitor (INH) effects (shown in units/mL) on Il-1 stimulated synthesis of PGE_2 in fibroblasts. Each point represents the mean ± SEM of two separate experiments. (Reprinted with permission from Korn JH, Brown KM, Downie E, Liao ZH, Rosenstreich DL: Augmentation of Il-1 induced fibroblast PGE_2 production by a urine-derived Il-1 inhibitor. *J. Immunol.* 138:3290–4, 1987.)

responses, suggesting a direct modulation of the PGE_2–IL-1 relationship in the disease process.

Where PGE_2 inhibits, LT appears to stimulate lymphokine activity. Goodwin et al. showed that LTB_4 production is a necessary component of mitogen-stimulated IL-2 production and T-lymphocyte proliferation (84). LTB_4 is normally synthesized during the first 24 h after exposure of T cells to PHA. If this is blocked by hydrocortisone or lipoxygenase inhibitors, both IL-2 production and T-lymphocyte proliferation is inhibited (84). Lymphocyte response can be restored by the addition of exogenous LTB_4, leading to the conclusion that LTB_4 plays a role in T-cell proliferation through stimulation of IL-2 production. This interpretation is supported by studies showing that LTB_4 induces proliferation of an IL-2-dependent T cell line (HT-2) in the presence of suboptimal concentrations of IL-2 (85). Representative results are shown in

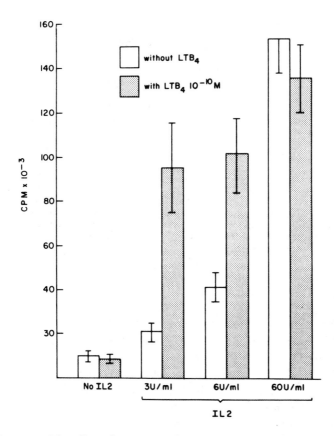

Figure 4.8. The effect of LTB$_4$ (10^{-10} M) on [^{3}H]thymidine incorporation in HT-2 cells cultured with various concentrations of Il-2. Data represent the mean ± SEM of five experiments. (Reprinted with permission from Atluru D, Goodwin JS: Leukotriene B$_4$ causes proliferation of interleukin 2-dependent T cells in the presence of suboptimal levels of interleukin 2. *Cell. Immunol.* 99:444–52, 1986.)

Figure 4.8. LTB$_4$ has little effect on cells cultured without IL-2, or with optimal IL-2 added to the medium. Thus, LTB$_4$ at physiologic concentrations (10^{-10} M) can substitute for IL-2 in the stimulation of IL-2-responsive T lymphocytes.

IL-2 is required for the production of interferon-$_\gamma$ (IFN$_\gamma$) by T lymphocytes (86). It also appears that leukotrienes (LTB$_4$, LTC$_4$, or LTD$_4$) can replace IL-2 in this inducer capacity (87). Johnson et al. examined the structural basis for the inducer signal for IFN$_\gamma$ production using arachidonic acid analogs and 5- and 15-HETE, as well as LTC$_4$ in the induction

Table 4.9. *The ability of LTC$_4$, arachidonic acid, and 5- and 15-HETE to provide helper signals for IFN$_\gamma$ production by Lyt-1(−) 2(+) cells.*

Fatty Acid	Concentration for Maximal Help (μM)	Percent of Control
None	—	<3
LTC$_4$	0.008	118
Arachidonic acid	0.003	171
5-HETE	0.015	161
15-HETE	0.015	81

^{a} Spleen cells were treated with monoclonal anti-Lyt-1.2 plus complement for 1 hr at 37°C, and the resulting helper cell-depleted cultures were stimulated for IFN-γ production with 0.5 μg SEA/ml in 2 or 3 day cultures. Control cultures were reconstituted with 60 μM dcGMP. IFN-γ activity of such cultures ranged from 93 to 140 U/ml. The data are representative of a minimum of three titrations.

Source: Reprinted with permission from Johnson HM, Russell JK, Torres BA: Second messenger role of arachidonic acid and its metabolites in interferon-production. *J. Immunol.* 137:3053–6, 1986.

of IFN$_\gamma$ (Table 4.9) (88). Johnson et al. concluded from their studies that arachidonic acid and its lipoxygenase products play a central role as helper signals for IFN$_\gamma$ production.

More recently, Rola-Pleszczynski et al. have shown that human T cells, pulsed with LTB$_4$, modulate IL-1 production by human monocytes by secreting IFN$_\gamma$ (89). They demonstrated that LTB$_4$-pulsed T cells were capable of suppressing lymphocyte proliferation, if they were able to interact with monocytes, and that this suppressive effect could be reversed by the addition of indomethacin (89).

Prostaglandin-associated suppressor factors

At least passing reference should be made to several reports of PG-associated substances, which appear to participate in the expression of immunological activity. Rogers et al. reported that glass-adherent murine splenic T lymphocytes, cultured in the presence of PGE$_2$ (10^{-5} M), produced a substance capable of suppressing PHA- and LPS-induced blastogenesis (90). The suppressor, given the acronym PITS (for prostaglandin-induced T-cell-derived suppressor), was found to be DNase, RNase, and heat resistant, but sensitive to treatment with proteinase K, trypsin, and Pronase. Further characterization showed two active moieties with approximate molecular weights of 35 000 daltons (PITS$_\alpha$) and 5000 daltons (PITS$_\beta$). Results of experiments with cycloheximide-treated glass-adherent T lymphocytes suggested that PGE_2 functions by inducing the release of PITS rather than its synthesis (90).

Likewise, Ninnemann and Stockland suggested that a low molecular weight carrier may be induced by and involved in the expression of PGE-mediated lymphocyte suppression following thermal injury (91). Further work toward characterization of this putative carrier has led to the definition of an injury-associated peptide, named suppressor active peptide (SAP) (92,93). Recent results obtained by Ozkan et al. indicate that SAP may function as an inducer of PGE_2 synthesis and act in concert with released PG in the expression of reduced immune function (94).

Finally, Kato and Askenase have reported the existence of T-cell products that can function as carriers for PGE (95). The system they employed was the trinitro phenol (TNP)-suppressor factor model in which suppressor T cells release a substance, which mediates picrylchloride contact sensitivity (93,94). Incubation of suppressor-factor-producing cells in the presence of indomethacin blocked the production of active suppressor factor. Instead, an inactive molecule was generated, which could be activated by the addition of PGE_1 or PGE_2, suggesting that in addition to mediating nonspecific immunosuppression, PGs play a role in antigen-specific immunoregulation (96).

ARACHIDONIC ACID METABOLISM AND B-CELL FUNCTION

The possibility that PGs are involved in the regulation of B-cell function was introduced by Webb and Osheroff, who demonstrated that mice, pretreated with indomethacin, had an improved plaque-forming cell response after sheep erythrocyte (sRBC) immunization (97). Zimecki and Webb also studied the response of mouse splenocytes *in vitro* to polyvinyl pyrolidone (PVP) and dinitrophenol (DNP)-Ficoll, both T-independent antigens (98). Plaque-forming cell responses were increased 50–300% in the presence of PGE, even when T cells and macrophages were specifically depleted. They concluded from these results that B cells might be capable of regulating their own response to certain antigens through the production of PGs (98). Like their murine counterpart, human B cells are also affected by PGs, specifically PGE_2.

B-cell suppression by exogenous PGE_2

Thompson et al. reported PGE_2 suppression of DNA synthesis in highly purified human peripheral blood B cells, after their stimulation with

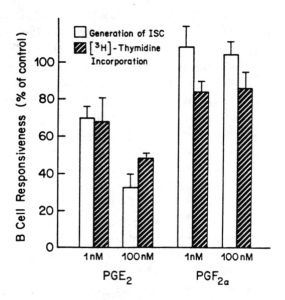

Figure 4.9. Differential effects of PGE_2 and $PGF_{2\alpha}$ on the responses of human B cells. B cells (2.5×10^4 per well) were cultured with *Staphylococcus aureus* and T cell supernatant and assayed for the generation of immunoglobulin-secreting cells (ISC) and [³H]thymidine incorporation after a 5-day incubation. Responses are given as mean percentage of control values obtained from three experiments. (Reprinted with permission from Simkin NJ, Jelinek DF, Lipsky PE: Incubation of human B cell responsiveness by prostaglandin E_2. *J. Immunol.* 138:1074–81, 1987.)

Staphylococcus aureus (99). These results were not altered by the presence or absence of T-cell supernatants containing B-cell growth factors. PGE_2-mediated suppression of stimulated B cells was clearly concentration dependent; significant inhibition was observed at PGE_2 concentrations as low as 10^{-9} M (99). Jelinek et al. studied the role of PGE_2 in human B-cell differentiation into mature antibody-forming cells *in vitro* (100) and found that PGE_2 inhibited the generation of immunoglobulin-secreting cells in response to *Staphylococcus aureus* or to T-cell supernatants containing B-cell differentiation factor. Recent studies by Simkin et al., utilizing carefully isolated and purified human B cells, confirmed the ability of PGE_2 to suppress the clonal expansion of activated cells, while $PGF_{2\alpha}$ was found to have no effect (Figure 4.9) (101).

There appear to be differences, however, in the way human and murine B cells are affected by PGE_2. Simkin et al. showed that in human

Table 4.10. *The results of two experiments showing the inhibition of the generation of Ig-secreting cells (ISC) by the addition of PGE_2 or forskolin to cultures.*[a]

Additions	Generation of ISC (ISC per 10^6 B cells × 10^{-3})	
	S. aureus	S. aureus + T Supt
Experiment 1		
0	0	66.8 ± 3.2
PGE_2	0	11.0 ± 0.7
Forskolin	0	14.4 ± 0.9
Experiment 2		
0	0	105.4 ± 9.5
PGE_2	0	30.2 ± 1.5
Forskolin	0	20.2 ± 0.8

[a] Peripheral blood B cells were incubated with *S. aureus*, or with the combination of *S. aureus* + T supt in the presence or absence of either PGE_2 or forskolin and assayed for ISC generation after 5 days.

Source: Reprinted with permission from Simkin NJ, Jelinek DF, Lipsky PE: Inhibition of human B cell responsiveness by prostaglandin E_2. *J. Immunol.* 138:1074–81, 1987.

B cells PGE_2 suppressed DNA synthesis and the generation of immunoglobulin-secreting cells at 10^{-9} M concentrations (101); in the mouse, higher concentrations are generally required (97,102). Ono et al. reported that in the mouse, only B-cell responses supported by a specific T-cell factor ($B151$-TRF_2) were suppressed by PGE_2 (102), while human B-cell responses supported by either crude T-cell supernatants or IL-2 were suppressed (101).

As PGE_2 increases intracellular cyclic AMP concentrations and $PGF_{2\alpha}$ does not, the differential effect of these metabolites on B-cell response supports the hypothesis that PGE_2 suppression is direct and results from the elevation of this cyclic nucleotide (101,103). Simkin et al. utilized a known intracellular cyclic AMP stimulator, forskolin, to test the effect of elevating cAMP on B-cell response (101). Results indicated that the effects of PGE_2 and forskolin on B-cell function appeared to be comparable (Table 4.10), at least circumstantially implicating cyclic AMP. It also appeared that PGE_2 had no significant effect on B-cell DNA synthesis during the first 60 h of incubation *in vitro* (101). Exogenous PGE_2, therefore, does not appear to inhibit initial B-cell activation, but rather appears to suppress ongoing proliferation and differentiation of B cells after initial activation and cell division.

Jelinek et al. have reported that B-cell susceptibility to PGE_2 inhibition is increased when B cells are stimulated with polyclonal activators, which cross-link surface immunoglobulin (100). Anti-Ig and other cross-linkers of surface Ig have been shown to increase intracellular calcium $[Ca^{2+}]$ concentrations in B cells (104,105), and it appears that this mechanism accounts for their increased PGE_2 susceptibility (101).

Glucocorticoid-induced immunoglobulin production

In 1981, Grayson et al. reported that glucocorticosteroids were capable of inducing polyclonal Ig production in cultures of human peripheral blood mononuclear cells (106). The mechanism by which glucocorticosteroids stimulate B cells in this and other systems is unknown; however, several research groups have reported that glucocorticoids inhibit the release of arachidonic acid metabolites (107,108). This effect appears to be caused by the formation of a phospholipase-A_2-inhibiting glycoprotein (macrocortin or lipomodulin) by leukocytes exposed to glucocorticoids (109,110). It is thought that, by these mechanisms, infusions of corticosteroids prevent the release of metabolites or arachidonic acid in isolated, perfused organs such as the lung (107).

Goodwin and Atluru found that nonspecific inhibitors of lipoxygenase/cyclooxygenase stimulate polyclonal Ig production in a manner similar to the effect of glucocorticoids (Figure 4.10) (109). Specific cyclooxygenase inhibitors, on the other hand, inhibited Ig production. Finally, they found that two specific 5-lipoxygenase inhibitors stimulated Ig production, an effect which could be reversed by the addition to cultures of LTB_4 in low concentrations (10^{-10} M) on days one, two, and three. LTB_4 additions also reversed the stimulatory effect of glucocorticoids on such cultures, suggesting the importance of this metabolite in glucocorticoid stimulation of Ig production (111).

Suppressor cells and immunoglobulin synthesis

That PGE_2 may also have a direct positive effect on the response of human peripheral blood Ig-producing cells is suggested by the work of Ceuppens and Goodwin (112). Using isolated, PWM-stimulated lymphocytes, it was found that the administration of three different cyclooxygenase inhibitors (indomethacin, carprofen, and piroxicam) resulted in suppressed Ig synthesis (Figure 4.11). The suppression could be reversed by the addition of low doses (3×10^{-9}–3×10^{-8} M) PGE_2 as shown in Figure 4.12 (112). The effects of indomethacin and PGE_2 were eliminated when T cells were removed or inactivated prior to culture, and it was found that endogenenously produced PGE exerts a tonic inhibitory effect

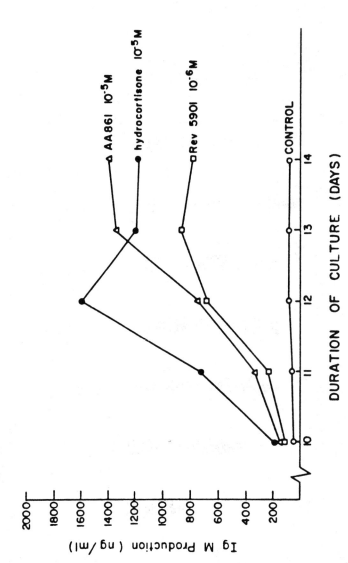

Figure 4.10. The effect of hydrocortisone and the specific 5-lipoxygenase inhibitors (AA861 and REV 5901) on polyclonal IgM production in cultures of peripheral blood mononuclear cells. Data are from a representative experiment. (Reprinted with permission from Goodwin JS, Atluru D: Mechanism of action of glucocorticoid-induced immunoglobulin production; role of lipoxygenase metabolites of arachidonic acid. *J. Immunol.* 136:3455–60, 1986.)

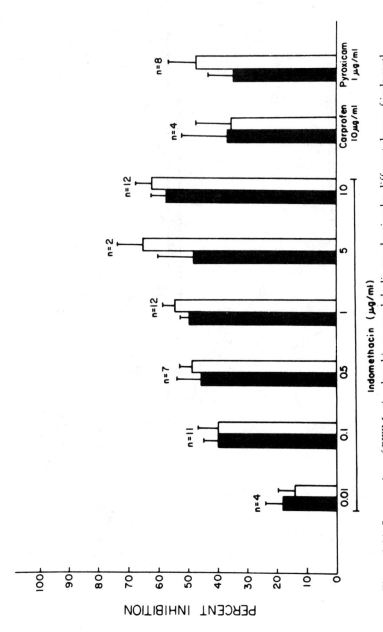

Figure 4.11. Suppression of PWM-stimulated immunoglobulin production by different doses of indomethacin and by two other cyclooxygenase inhibitors, piroxicam and carprofen. Results are expressed as percentage inhibition (mean ± SEM) in comparison to control cultures without drugs. Filled bars represent the results of IgG and the open bars for IgM. The number of experiments performed is indicated above the bars. (Reprinted with permission from Ceuppens JL, Goodwin JS: Endogenous prostaglandin E_2 enhances polyclonal immunoglobulin production by tonically inhibiting T-suppressor cell activity. *Cell. Immunol.* 70:41–54, 1982.)

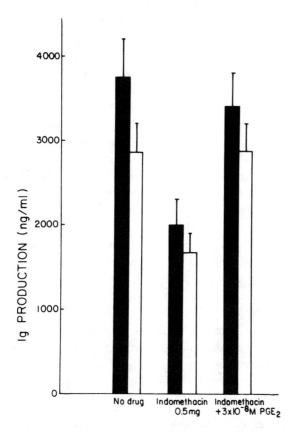

Figure 4.12. The effect of 0.5 μ/ml indomethacin, and of indomethacin plus 3 × 10⁻⁸ M PGE₂ on IgM (open bars) and IgG (filled bars) production. Results are expressed in nanograms per milliliter IgM or IgG and represent the mean ± SEM of 29 experiments. Similar results were obtained with two other concentrations of pokeweed mitogen. Indomethacin significantly decreased IgG ($p < 0.0001$) and IgM ($p < 0.0001$) production by paired t-test. The addition of PGE eliminated this difference. (Reprinted with permission from Ceuppens JL, Goodwin JS: Endogenous prostaglandin E₂ enhances polyclonal immunoglobulin production by tonically inhibiting T suppressor cell activity. *Cell. Immunol.* 70:41–54, 1982.)

on a radiosensitive, suppressor T cell, primarily, but not entirely of the OKT8(+) T-cell subset (112).

    The enhancement of Ig production by endogenous PGE₂ through the inhibition of a radiosensitive suppressor cell compartment has some important implications. In several diseases with autoantibody production, a deficiency of suppressor T cells is thought to be responsible for B-cell hyperreactivity. This is discussed in Chapter 8. Also, aging has been associated with decreased suppressor cell function, and the data generated by Ceuppens and Goodwin suggest that one explanation for this depressed function is an increased sensitivity to inhibition by PGE_2, an endogenous regulator of suppressor cell activity (112, 113).

Depending on their quantity and nature, the presence of arachidonic acid metabolites may also induce the activity of immunoregulatory cells, detrimental to B-cell function. Atluru and Goodwin have reported that LTB_4 is a potent suppressor of polyclonal Ig production in PWM-stimulated cultures of human peripheral blood lymphocytes while LTC_4 and LTD_4 have little activity in this system (Figure 4.13) (71). LTB_4 suppression of Ig synthesis was mediated by the induction of an OKT8(+) suppressor cell from an OKT8(−) subpopulation after 18-h incubation. The activity of these cells could be induced with as little as 10^{-10} M LTB_4, and could be blocked by cycloheximide but not by mitomycin C (71).

Finally, not all of the immunoregulatory cells induced by arachidonic acid metabolites appear to be T cells. Using a plaque-forming cell-response assay with sheep erythrocytes as the antigen in $C_{57}BL/6$ mice, Shimamura et al. have presented evidence suggesting that endogenous PG synthesis is responsible for the generation of suppressor-inducer B lymphocytes, affecting the response to high-dose sRBC (114).

B-cell tolerance and prostaglandin E_2

Studies on the induction of B-cell tolerance have included examination of the susceptibility of antigen-specific lymphocytes to being rendered unresponsive, and more recently, the role of the macrophage in tolerance induction and maintenance. The macrophage appears to play a key role, and in various systems has been reported to prevent (115), facilitate (116), and break (117) tolerance. Since, as noted in the previous chapter, macrophages are capable of releasing a wide repertoire of biochemical products with immunoregulatory properties (including PGE_2), it is possible that macrophages modulate tolerance induction at the B-cell level through the release of soluble mediators. Indeed, Goldings has reported that the development of B-cell tolerance is dependent upon macrophage production of PGE_2 (118). Indomethacin and acetylsalicylic acid were capable of blocking hapten-specific B-cell tolerance induction to the hapten, trinitrophenyl-human gammaglobulin. Tolerance could be completely restored by the addition of 6×10^{-9} M PGE_2. Tolerance induc-

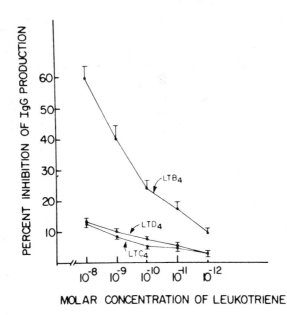

MOLAR CONCENTRATION OF LEUKOTRIENE

Figure 4.13. IgG production in pokeweed-mitogen-stimulated cultures of human peripheral blood mononuclear cells supplemented with LTB_4, LTC_4, or LTD_4. IgG was measured by enzyme linked immunosorbant assay (ELISA) after 7 days incubation. LTB_4 at all concentrations $(10^{-8}–10^{-12}$ M) caused a significant inhibition of IgG production ($p <$ 0.01 by paired t-test). High concentrations of LTC_4 and LTD_4 caused a small but significant inhibition of IgG production. Data represent mean $\pm$ SEM of the results of experiments on seven subjects. (Reprinted with permission from Atluru D, Goodwin JS: Control of polyclonal immunoglobulin production from human lymphocytes by leukotrienes; leukotriene B_4 induces an OKT8(+), radiosensitive suppressor cells from resting, human OKT8(−) T cells. *J. Clin. Invest.* 74:1444–50, 1984.)

tion was also blocked by macrophage IL-1 production, reflecting the control function of the macrophage through the elution of IL-1 or PGE_2 (118).

ARACHIDONIC ACID METABOLISM AND NATURAL KILLER
CELL FUNCTION

Natural killer (NK) cells are a subpopulation of lymphoid cells, present in a wide range of mammalian and avian species, which have spontaneous cytolytic activity against a variety of cells (119). NK cells originate in the fetal liver and are present in human cord blood at birth (120). Following birth, there is an increase in NK activity to adult levels, with NK cells found in peripheral blood and spleen, bone marrow, lymph nodes, and thoracic duct (121). Human NK cells are large granular lymphocytes (LGL) with a high cytoplasm:nucleus ratio, which bear receptors for the Fc portion of IgG, and have low-affinity receptors for sheep erythrocytes (122). Two monoclonal antibodies have been shown to bind selectively to NK cells. Almost all (>92%) LGL are HNK-1(+) cells, which have the characteristics of NK cells (123), whereas 33% of LGL are NK-8(+) (124). It is thought that HNK-1 binds to all LGL, and NK-8 binds to only those cells capable of interacting with target cells. NK cells account for approximately 5% of peripheral blood or splenic leukocytes (125). NK cells appear to be able to mediate natural resistance against tumors *in vivo*, certain virus and other microbial diseases, and transplanted tissue such as bone marrow (reviewed in 126). It now appears that the products of arachidonic acid metabolism may be involved in the regulation of these cells *in vivo*.

Because NK-cell cytotoxicity may be an important component of cancer resistance, Bankhurst initiated a study to determine whether the concentrations of PGs, produced *in situ* by tumors, was sufficient to depress NK activity (127). Using a standard assay measuring ^{51}Cr release from labeled K562 target cells, he found that NK cells were profoundly inhibited by high concentrations $(10^{-6}M)$ of PGE_2, PGD_2, PGA_2, and $PGF_{2\alpha}$. At physiological concentrations $(10^{-8}$ M$)$, however, only PGE_2 and PGD_2 were significantly suppressive. Bankhurst determined that PG effects were indeed directed at the effector cell (rather than target cell), and that supernatants from various tumor cell lines were also suppressive, in proportion to their PG content (127). Jondal et al. (128) showed that monocytes controlled NK-cell activity by a similar mechanism. Using ^{51}Cr release from MOLT-4 target cells as the standard assay, monocyte-induced suppression of NK cytotoxicity was found to be independent of cell–cell contact, and could be blocked by the presence of indomethacin (128).

Droller et al. showed that drugs such as indomethacin, fenelozic acid, acetylsalicylic acid, and 2-6-xylenol, known to inhibit PG production by tumor target cells, were capable of enhancing both NK and antibody-dependent cellular cytotoxicity (ADCC) (129). Droller et al. also demonstrated that exogenous PGE_1, PGE_2 and theophylline partially inhibited

NK activity and ADCC, and hypothesized that this alteration was caused by increases in cyclic AMP levels in the attacking lymphocytes (130). While PGs inhibit, it appears that another monocyte product, interferon (IFN), can augment NK cytotoxicity. The activity of NK cells has been observed to be more greatly enhanced by IFN, and more greatly depressed by PGE_1, than other forms of cytotoxicity such as ADCC (131). Also, incubation *in vitro* of human NK cells with IFN results in a partial loss of their sensitivity to inhibition by PGE_2, an effect which may also be important to tumor resistance *in vivo* (132).

Where PGD and PGE serve as "off" signals for NK activity, lipoxygenase products may serve as "on" signals (133–135). As shown in Figure 4.14, nordihydroguaiaretic acid (NDGA), quercetin, eicosatetraynoic acid (ETYA), phenidone, and esculetin (all agents known to inhibit cellular lipoxygenase activity) also inhibit human NK cytotoxicity in a dose-dependent manner (135). The 5-lipoxygenase products 5-HPETE and LTB_4 were found to enhance NK activity significantly, although inhibition of LTB_4 by diethylcarbamizine did not diminish NK activity versus K562 tumor target cells *in vitro*. Most significantly, these studies by Bray and Brahmi (135) and studies by Seaman and Woodcock (134) indicated that lipoxygenase activity not only enhanced, but also appeared to be required for an NK-cytotoxic response.

Serhan et al. studied the effect of lipoxins on NK-cell function and found that concentrations of approximately 10^{-7} M of lipoxin A (LXA) or lipoxin B (LXB) produced 50% inhibition of K562 target cell lysis (136). The inhibition of NK activity by synthetic LXB and its isomers is shown in Figure 4.15. Neither LXA or LXB elevated intracellular cyclic AMP, nor inhibited target-cell binding, which suggested a mode of action, such as activation of protein kinase, which is not related to target–effector-cell recognition. Serhan et al. also reported that 15-HETE, LTB_4 and LTC_4 were unable to inhibit NK lysis of K562 target cells, suggesting that the presence of a trihydroxytetraene structure (such as found in the lipoxins) is necessary for inhibition to occur (136).

Lymphocyte sensitivity changes

A multitude of variables is known to effect the response of lymphocytes to the presence of arachidonic acid metabolites. Staszak et al., for example, have suggested that the major histocompatibility locus plays a role in PG sensitivity in human lymphocytes, and have presented evidence that

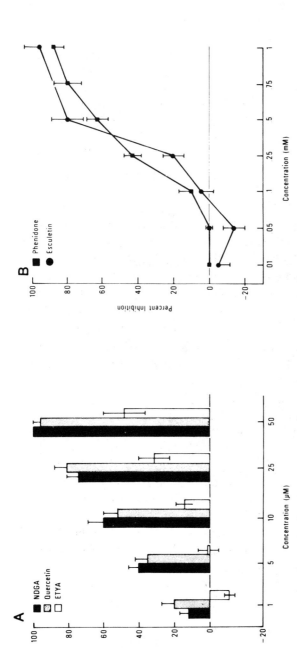

Figure 4.14. The effects of various lipoxygenase inhibitors on NK cytotoxicity. (A) Inhibition of NK activity when NDGA, quercetin, and ETYA were added directly to the ⁵¹Cr-release assay. Results represent the mean ± SEM of 10 experiments at an effector cell:target cell ratio of 20:1. Similar results were observed at a 10:1 ratio. (B) Inhibition of NK activity when phenidone and esculetin were added directly to the ⁵¹Cr-release assay. Results represent the mean ± SEM of 6 experiments at an EC:TC ratio of 20:1. Similar results were observed at a 10:1 ratio. (Reprinted with permission from Bray RA, Brahmi Z: Role of lipoxygenation in human natural killer cell activation. *J. Immunol.* 136:1783–90, 1986.)

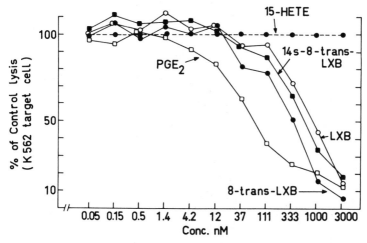

Figure 4.15. The inhibition of human NK cytotoxicity versus K562 target cells by lipoxin B and its isomers, PGE₂ (positive control) and 15-HETE (negative control). (Reprinted with permission from Serhan CN, Hamberg M, Ramstedt U, Samuelsson B: Lipoxins, stereochemistry, biosynthesis, and biological activities. *Adv. Prostagl. Thrombox. Leukotr. Res.* 16:83–97, 1986.)

the HLA-B12 haplotype, often associated with autoimmune disease is also associated with decreased sensitivity to the immunoinhibitory effects of PGE₂ (137). Conversely, lymphocytes from human subjects subjected to stress (childbirth or cardiac surgery) were more sensitive to inhibition by PGE₂ (138), as are lymphocytes subjected to the stress of major injuries (139). This increased sensitivity was accompanied by significantly decreased lymphocyte proliferation *in vitro* (17–68% of controls) and a 50% increase in E-rosette-positive cells with receptors for the Fc portion of IgG (138). This effect may or may not be related to the known synergistic activity of cortisol and PGE₂ in suppressing lymphocyte response (140). An increase in the sensitivity of lymphocytes to the effects of PGE has also been noted to be a consequence of the aging process (113).

Rocklin et al. showed that mononuclear cells derived from atopic individuals were less sensitive than normal to the suppressive effects of PGD₂ and PGE₂, with regard to protein synthesis and activation of suppressor cells (141). Changes in lymphocyte sensitivity to the products of arachidonic acid metabolism, therefore, may be important to the induction and onset of malignancy, transplant rejection, rheumatoid arthritis, immune depression following injury or surgery, as well as in the etiology of al-

88   *Prostaglandins, leukotrienes, and the immune response*

lergy. These possibilities are discussed in detail in the chapters which follow.

1. Parker CW, Stenson WF, Huber MG, Kelly JP: Formation of thromboxane B$_2$ and hydroxyarachidonic acids in purified lymphocytes in the presence and absence of PHA. *J. Immunol.* 122:1572–6, 1979.
2. Goetzl EJ: Selective feed-back inhibition of the 5-lipoxygenation of arachidonic acid in human T-lymphocytes. *Biochem. Biophys. Res. Commun.* 101:344–50, 1981.
3. Bauminger S: Differences in prostaglandin formation between thymocyte subpopulations. *Prostaglandins* 16:351–5, 1978.
4. Webb DR, Nowowiejski I: Mitogen-induced changes in lymphocyte prostaglandin levels: a signal for the induction of suppressor cell activity. *Cell. Immunol.* 41:72–8, 1978.
5. Aussel C, Didier M, Fehlmann M: Prostaglandin synthesis in human T cells: its partial inhibition by lectins and antiCD3 antibodies as a possible step in T cell activation. *J. Immunol.* 138:3094–9, 1987.
6. Kennedy MS, Stobo JD, Goldyne ME: In vitro synthesis of prostaglandins and related lipids by populations of human peripheral blood mononuclear cells. *Prostaglandins* 20:135–45, 1980.
7. Bankhurst AD, Hastain E, Goodwin JS, Peake GT: The nature of the prostaglandin-producing mononuclear cell in human peripheral blood. *J. Lab. Clin. Med.* 97:179–86, 1981.
8. Goldyne ME, Burrish GF, Poubelle P, Borgeat P: Arachidonic acid metabolism among human mononuclear leukocytes. *J. Biol. Chem.* 259:8815–19, 1984.
9. Goodwin JS, Ceuppens J: Regulation of the immune response by prostaglandins. *J. Clin. Immunol.* 3:295-315, 1983.
10. Kantor HS, Hampton M: Indomethacin in submicromolar concentrations inhibits cyclic AMP-dependent protein kinase. *Nature* (London) 276:841–4, 1978.
11. Ciosek CP, Ortel RW, Thanassi NM, Newcombe DS: Inhibition of phosphodiesterasse by nonsteroidal antiinflammatory drugs. *Nature* (London) 251:148–50, 1974.
12. Trang LE: Prostaglandins and inflammation. *Sem. Arth. Rheum.* 9:153–9, 1980.
13. Goodwin JS, Webb DR: Regulation of the immune response by prostaglandins. *Clin. Immunol. Immunopathol.* 15:106–22, 1980.
14. Fitzpatrick FA, Wynalda MA: In vivo suppression of prostaglandin biosynthesis by nonsteroidal antiinflammatory agents. *Prostaglandins* 12:1037–51, 1976.
15. Naden RP, Iliya CA, Arant BS, Grant NF, Rosenfled CR: Hemodynamic effects of indomethacin in chronically instrumented pregnant sheep. *Am. J. Obstet. Gynecol.* 1511:484–94, 1985.

16. Christ-Hazelhof E, Nugteren DH: Prostacyclin is not a circulating hormone. *Prostaglandins* 22:739–46, 1981.

17. Granström E, Samuelsson B: Radioimmunoassays for prostaglandins. *Adv. Prost. Thrombox. Res.* 5:1–53, 1978.

18. Gorman RR, Fitzpatrick FA, Miller OV: Reciprocal regulation of human platelet cAMP levels by thromboxane A_2 and prostacyclin. *Adv. Nucleotide Res.* 9:597–609, 1978.

19. Torikai S, Kurokawa K: Distribution of prostaglandin E_2-sensitive adenylate cyclase along the rat nephron. *Prostaglandins* 21:427–38, 1981.

20. Mexmain S, Cook J, Aldigier JC, Gualde N, Rigaud M: Thymocyte cyclic AMP and cyclic GMP responses to treatment with metabolites issued from the lipoxygenase pathway. *J. Immunol.* 135:1361–5, 1985.

21. Goodwin JS, Wiik A, Lewis M, Bankhurst AD, Williams RC: High affinity binding sites for prostaglandin E on human lymphocytes. *Cell. Immunol.* 43:150–9, 1979.

22. DeMeyts P, Roth J, Neville DM, Gavin JR, Lesniak MA: Insulin interactions with its receptors: experimental evidence for negative cooperativity. *Biochem. Biophys.Res. Commun.* 55:154–61, 1973.

23. Rao CV: Comparisons of affinity constants of PGEs: receptor interaction with concentration of PGEs needed for half maximum stimulation of adenylate cyclase. *Prostaglandins* 9:579–84, 1975.

24. Brunton LL, Wiklund RA, Van Arsdale PM, Gilman AG: Binding of [^{3}H]prostaglandin E_1 to putative receptors linked to adenylate cyclase of cultured cell clones. *J. Biol. Chem.* 251:3037–44, 1976.

25. Henney CS, Bourne HR, Lichtenstein LM: The role of cyclic $3',5'$ adenosine monophosphate in the specific cytolytic activity of lymphocytes. *J. Immunol.* 108:1526–34, 1972.

26. Bach MA: Differences in cyclic AMP changes after stimulation by prostaglandins and isoproterenol in lymphocyte subpopulations. *J. Clin. Invest.* 55:1074–81, 1975.

27. Zurier RB, Dore-Duffy P, Viola MV: Adherence of human peripheral blood lymphocytes to measles-infected cells. *N. Engl. J. Med.* 296:1443–6, 1977.

28. Sinha AK, Colman RW: Prostaglandin E_2 inhibits platelet aggregation by a pathway independent of adenosine $3',5'$monophosphate. *Science* 200:202–3, 1978.

29. Bito LZ: Saturable, energy dependent, transmembrane transport of prostaglandins against concentration gradients. *Nature* (London) 256:134–6, 1975.

30. Goodwin JS, Bromberg S, Messner RP: Studies on the cyclic AMP response to prostaglandin in human lymphocytes. *Cell. Immunol.* 60:298–307, 1981.

31. Smith JW, Steiner AL, Parker CW: Human lymphocyte metabolism: effects of cyclic and noncyclic nucleotides on stimulation by phytohemagglutinin. *J. Clin. Invest.* 50:442–8, 1971.

32. Koopman WJ, Gillis MH, David JR: Prevention of MIF activity by agents known to increase cellular cyclic AMP. *J. Immunol.* 110:1609–14, 1973.

33. Lomnitzer R, Rabson AR, Koornhof HJ: The effects of cyclic AMP on

leukocyte inhibitory factor production and on the inhibition of leukocyte migration. *Clin. Exp. Immunol.* 24:42–8, 1976.
34. Melmon KL, Bourine HR, Weinstein Y, Shearer GM, Keman J, Bauminger S: Hemolytic plaque formation by leukocytes in vitro. *J. Clin. Invest.* 53:13–21, 1974.
35. Braun W, Ishizuka M: Antibody formation: Reduced responses after administration of excessive amounts of nonspecific stimulators. *Proc. Nat. Acad. Sci. USA* 68:114–6, 1971.
36. Goodwin JS, Messner RP, Peake GT: Prostaglandin suppression of mitogen-stimulated lymphocytes in vitro. *J. Clin. Invest.* 62:753–60, 1978.
37. Novagrodsky A, Rubin AL, Stenzel KH: Selective suppression by adherent cells, prostaglandin, and cyclic AMP analogues of blastogenesis induced by different mitogens. *J. Immunol.* 122:1–8, 1979.
38. Eakles DD, Gershwin ME: Pharmacologic and biochemical production of human T lymphocyte colony formation: hormonal influences. *Immunopathology* 3:259–74, 1981.
39. Bockman RS: PGE inhibition of T lymphocyte colony formation. *J. Clin. Invest.* 64:812–21, 1979.
40. Gorski A, Graciong Z, Dupont B: Enhanced colonability of human T lymphocytes caused by their culturing in vitro. *Immunology* 44:617–22, 1982.
41. Enteu U, Enre T, Cavdar AO, Turker RK: An in vitro study on the effect of prostaglandin E_2 and F_2 on E-rosette forming activity of normal lymphocytes. *Prostagl. Med.* 5:255–61, 1980.
42. Venza-Teti D, Misefari A, Sofo V, Fimiani V, LaVia MF: Interaction between prostaglandins and human T lymphocytes: Effect of PGE_2 on E receptor expression. *Immunopharmacology* 2:165–71, 1980.
43. Gordon D, Bray M, Morley J: Control of lymphokine secretion by prostaglandins. *Nature* (London) 262:401–7, 1976.
44. Baker PE, Fahey JV, Munck A: Prostaglandin inhibition of T-cell proliferation is mediated at two levels. *Cell. Immunol.* 61:52–7, 1981.
45. Rappaport RS, Dodge GR: Prostaglandin E inhibits the production of human interleukin 2. *J. Exp. Med.* 155:943–8, 1982.
46. Sappi E, Eskola J, Ruskanen O: Effects of indomethacin on lymphocyte proliferation, suppressor cell function, and leukocyte migration inhibitory factor (LMIF) production. *Immunopharmacology* 4:236–42, 1982.
47. Darrow TL, Tomar RH: Prostaglandin-mediated regulation of the mixed lymphocyte culture and generation of cytotoxic cells. *Cell. Immunol.* 65:172–8, 1980.
48. Leung KH, Mihich E: Prostaglandin modulation of development of cell mediated immunity in culture. *Nature* (London) 288:597–603, 1980.
49. VanEpps D: Suppression of human lymphocyte migration by PGE_2. *Inflammation* 5:81–7, 1981.
50. Goodwin JS, Bankhurst AD, Messner RP: Suppression of human T-cell mitogenesis by prostaglandin: existence of a prostaglandin-producing suppressor cell. *J. Exp. Med.* 146:1719–34, 1977.
51. Goodwin JS, Messner RP, Williams RC: Inhibitors of T-cell mitogenesis: effect of mitogen dose. *Cell. Immunol.* 45:303–9, 1979.

52. Ninnemann JL, Stockland AE: Participation of prostaglandin E in immunosuppression following thermal injury. *J. Trauma* 24:201–7, 1984.

53. Goodman MG, Weigle WO: Modulation of lymphocyte activation: I. Inhibition by an oxidation product of arachidonic acid. *J. Immunol.* 125:593–600, 1980.

54. Gualde N, Rabinovitch H, Fredon M, Rigaud M: Effects of 15- hydroperoxyeicosatetraenoic acid on human lymphocyte sheep erythrocyte rosette formation and response to concanavalin A associated with HLA system. *Eur. J. Immunol.* 12:773–7, 1982.

55. Payan DG, Goetzl EJ: Specific suppression of human T lymphocyte function by leukotriene B_4. *J. Immunol.* 131:551–3, 1983.

56. Rola-Pleszczynski M, Borgeat P, Sirois P: Leukotriene B_4 induces human suppressor lymphocytes. *Biochem. Biophys. Res. Commun.* 108:1531–7, 1982.

57. Webb DR, Nowowiejski I, Healty C, Rogers TJ: Immunosuppressive properties of leukotriene D_4 and E_4 in vitro. *Biochem. Biophys. Res. Commun.* 104:1617–22, 1982.

58. Gualde N, Atluru D, Goodwin JS: Effect of lipoxygenase metabolites of arachidonic acid on proliferation of human T cells and T cell subsets. *J. Immunol.* 134:1125–9, 1985.

59. Stobo JD, Kennedy MS, Goldyne ME: Prostaglandin E modulation of the mitogenic response of human T cells. *J. Clin. Invest.* 64:1188–92, 1979.

60. Gualde N, Goodwin JS: Effects of prostaglandin E_2 and preincubation on lectin-stimulated proliferation of human T cell subsets. *Cell. Immunol.* 70:373–8, 1982.

61. Kendall R, Targen S: The dual effect of prostaglandin and ethanol on the natural killer cytolytic process. *J. Immunol.* 125:2770–7, 1980.

62. Mertin J, Stackpoole A: Anti-PGE antibodies inhibit in vivo development of cell-mediated immunity. *Nature* (London) 294:456–8, 1981.

63. Bach MA, Fournier C, Bach JF: Regulation of theta-antigen expression by agents altering cyclic AMP levels and by thymic factor. *Ann. N.Y. Acad. Sci.* 249:316–20, 1975.

64. Gualde N, Riguad M, Bach JF: Stimulation of prostaglandin synthesis by the serum thymic factor (FTS). *Cell. Immunol.* 70:362–7, 1982.

65. Rinaldi-Garaci C, Goblo V, Favalli C, Garaci E, Bistoni F, Jaffe B: Induction of serum thymic-like activity in adult thymectomized mice by a synthetic analog of PGE_2. *Cell. Immunol.* 72:97–101, 1982.

66. Fischer A, Durandy A, Griscelli C: Role of prostaglandin E_2 in the induction of nonspecific T lymphocyte suppressor activity. *J. Immunol.* 126:1452–5, 1981.

67. Kaszubowski PA, Goodwin JS: Monocyte-produced prostaglandin induces Fc_γ receptor expression on human T cells. *Cell. Immunol.* 68:343–8, 1982.

68. Rappaport RS, Dodge GR: Prostaglandin E inhibits the production of human interleukin 2. *J. Exp. Med.* 155:943–8, 1982.

69. Chouaib S, Chatenoud L, Klatzmann D, Fradelizi D: The mechanism of inhibition of human IL-2 production: II. PGE_2 induction of suppressor T lymphocytes. *J. Immunol.* 132:1851–7, 1984.

70. Fulton AM, Levy JG: The induction of nonspecific T suppressor lymphocytes by prostaglandin E$_1$. *Cell. Immunol.* 59:54–60, 1981.
71. Atluru D, Goodwin JS: Control of polyclonal immunoglobulin production from human lymphocytes by leukotrienes: leukotriene B$_4$ induces an OKT8(+), radiosensitive suppressor cells from resting, OKT8(−) T cells. *J. Clin. Invest.* 74:1444–50, 1984.
72. Gualde N, Goodwin JS: Induction of suppressor T cells in vitro by arachidonic acid metabolites issued from the lipoxygenase pathway. *In*: Bailey JM (ed.), *Prostaglandins, Leukotrienes, and Lipoxins*. Plenum Press, N.Y., pp. 565–75, 1983.
73. Aldigier JC, Gualde N, Mexmain S, Chable-Rabinovitch H, Ratinaud MH, Rigaud M: Immunosuppression induced in vivo by 15-hydroxyeicosatetraenoic acid (15 HETE). *Prostagl. Med.* 13:99–107, 1984.
74. Mexmain S, Gualde N, Aldigier JC, Motta C, Chable-Rabinovitch H, Rigaud M: Specific binding of 15-HETE to lymphocytes: effects on the fluidity of plasmatic membranes. *Prostagl. Med.* 13:93–7, 1984.
75. Gualde N, Rigaud M, Goodwin JS: Induction of suppressor cells from human peripheral blood T cells by 15-hydroperoxyeicosatetraenoic acid (15-HPETE). *Clin. Res.* 31:490A, 1983.
76. Chouaib S, Fradelizi D: The mechanism of inhibition of human IL-2 production. *J. Immunol.* 129:2463–8, 1982.
77. Tilden AB, Balch CM: A comparison of PGE$_2$ effects on human suppressor cell function and on interleukin 2 function. *J. Immunol.* 129:2469–73, 1982.
78. Gordon D, Bray MA, Morley J: Control of lymphokine secretion by prostaglandins. *Nature* (London) 262:401–2, 1976.
79. Baker PE, Fahey JV, Munck A: Prostaglandin inhibition of T cell proliferation is mediated at two levels. *Cell. Immunol.* 61:52–61, 1981.
80. Teodorczyk-Injeyan JA, Sparkes BG, Peters WJ, Gerry K, Falk RE: Prostaglandin E-related impaired expression of interleukin 2 receptor in the burn patient. *Adv. Prostagl. Thrombox. Leukotr. Res.* 17:151–4, 1987.
81. Ninnemann JL: PGE regulates lymphocyte stimulation by IL-2. *Proc. Am. Burn Assoc.* 19:10, 1987.
82. Kunkel SL, Chensue SW, Phan SH: Prostaglandins as endogenous mediators of interleukin 1 production. *J. Immunol.* 136:186–92, 1986.
83. Korn JH, Brown KM, Downie E, Liao ZH, Rosenstreich DL: Augmentation of IL 1-induced fibroblast PGE$_2$ production by a urine-derived IL 1 inhibitor. *J. Immunol.* 138:3290–4, 1987.
84. Goodwin JS, Atluru D, Sierakowski S, Lianois E: Mechanism of action of glycocorticosteroids. Inhibition of T cell proliferation and interleukin 2 production by hydrocortisone is reversed by leukotriene B4. *J. Clin. Invest.* 77:1244–50, 1986.
85. Atluru D, Goodwin JS: Leukotriene B$_4$ causes proliferation of interleukin 2-dependent T cells in the presence of suboptimal levels of interleukin 2. *Cell. Immunol.* 99:444–52, 1986.
86. Torres BA, Farrar WL, Johnson HM: Interleukin 2 regulates immune interferon (IFN$_\gamma$) production by normal and suppressor cell cultures. *J. Immunol.* 128:2217–19, 1982.

87. Johnson HM, Torres BA: Leukotrienes: positive signals for regulation of γ-interferon production. *J. Immunol.* 132:413–16, 1984.

88. Johnson HM, Russell JK, Torres BA: Second messenger role of arachidonic acid and its metabolites in interferon-production. *J. Immunol.* 137:3053–6, 1986.

89. Rola-Pleszczynski M, Bouvrette L, Gingras D, Girard M: Identification of interferon-gamma as the lymphokine that mediates leukotriene B₄-induced immunoregulation. *J. Immunol.* 139:513–17, 1987.

90. Rogers TJ, Nowowiejski I, Webb DR: Partial characterization of a prostaglandin-induced suppressor factor. *Cell. Immunol.* 50:82–93, 1980.

91. Ninnemann JL, Stockland AE: Participation of prostaglandin E in immunosuppression following thermal injury. *J. Trauma* 24:201–7, 1984.

92. Ozkan AN, Ninnemann JL: Suppression of in vitro lymphocyte and neutrophil responses by a low molecular weight suppressor active peptide from burn patient sera. *J. Clin. Immunol.* 5:172–9, 1985.

93. Ozkan AN, Ninnemann JL, Sullivan JJ: Progress in the characterization of an immunosuppressive glycopeptide (SAP) from patients with major thermal injuries. *J. Burn Care Rehabil.* 7:388–97, 1986.

94. Ozkan AN, Hoyt DB, Ninnemann JL, Mitchell MD: Suppressor active peptide-mediated prostaglandin E biosynthesis: a potential mechanism for trauma-induced immunosuppression. *Immunol. Lett.* (in press), 1987.

95. Kato K, Askenase PW: Reconstitution of an inactive antigen-specific T cell suppressor factor by incubation of the factor with prostaglandins. *J. Immunol.* 133:2025–31, 1984.

96. Rosenstein RW, Murray JH, Cone RE, Ptak W, Iverson GM, Gershon RK: Isolation and partial characterization of an antigen-specific T-cell factor associated with the suppression of delayed-type hypersensitivity. *Proc. Natl. Acad. Sci. USA* 78:5821–5, 1981.

97. Webb DR, Osheroff PL: Antigen stimulation of prostaglandin synthesis and control of immune responses. *Proc. Natl. Acad. Sci. USA* 73:1300–4, 1975.

98. Zimecki M, Webb DR: The regulation of the immune response to T-independent antigens by prostaglandins and B cells. *J. Immunol.* 117:2158–64, 1976.

99. Thompson PA, Jelinek DF, Lipsky PE: Regulation of human B cell proliferation by prostaglandin E₂. *J. Immunol.* 133:2446–53, 1984.

100. Jelinek DF, Thompson PA, Lipsky PE: Regulation of human B cell activation by prostaglandin E₂: suppression of the generation of immunoglobulin-secreting cells. *J. Clin. Invest.* 75:1339–49, 1985.

101. Simkin NJ, Jelinek DF, Lipsky PE: Inhibition of human B cell responsiveness by prostaglandin E₂. *J. Immunol.* 138:1074-81, 1987.

102. Ono S, Hayashi SI, Takahama Y, Bobashi K, Katoh Y, Nakanishi K, Paul WE, Hamaoka T: Identification of two distinct factors, B151-TRF1 and B151-TRF2, inducing differentiation of activated B cells and small resting B cells into antibody-producing cells. *J. Immunol.* 137:187–200, 1986.

103. Goodwin JS, Bromberg S, Messner RP: Studies on the cyclic AMP response to prostaglandins in human lymphocytes. *Cell. Immunol.* 60:298–307, 1981.

104. Braun J, Sha'afi I, Unanue ER: Cross-linking by ligands to surface immu-

noglobulin triggers mobilization of intracellular $^{42}Ca^{++}$ in B lymphocytes. *J. Cell. Biol.* 82:755–66, 1979.

105. Coggeshall KM, Cambier JC: B cell activation. VIII. Membrane immunoglobulins transduce signals via activation of phosphatidyl-inositol hydrolysis. *J. Immunol.* 133:3382–6, 1984.

106. Grayson J, Dooley N, Koski I, Blaese RM: Immunoglobulin production induced in vitro by glucocorticosteroid hormones. *J. Clin. Invest.* 68:1539–47, 1981.

107. Floman Y, Zor U: Mechanism of steroid action in inflammation: inhibition of prostaglandin synthesis and release. *Prostaglandins* 12:403–14, 1976.

108. Hong SL, Levine L: Inhibition of arachidonic acid release from cells as the biochemical action of antiinflammatory corticosteroids. *Proc. Natl. Acad. Sci. USA* 73:1730–4, 1976.

109. Blackwell GJ, Carnuccio R, DiRose M, Flover RJ, Paraente L, Perrico P: Macrocortin: A polypeptide causing anti phospholipase effect of glyucocorticoids. *Nature* (London) 287:147-9, 1980.

110. Hirata F, Schiffman E, Yenkatasubramanian K, Salomen D, Axelrod J: A phospholipse A_2 inhibitory protein in rabbit neutrophils induced by glucocorticoids. *Proc. Natl. Acad. Sci. USA* 77:2533–6, 1981.

111. Goodwin JS, Atluru D: Mechanism of action of glucocorticoid-induced immunoglobulin production: role of lipoxygenase metabolites of arachidonic acid. *J. Immunol.* 136:3455–60, 1986.

112. Ceuppens JL, Goodwin JS: Endogenous prostaglandin E_2 enhances polyclonal immunoglobulin production by tonically inhibiting T suppressor cell activity. *Cell. Immunol.* 70:41–54, 1982.

113. Goodwin JS: Changes in lymphocyte sensitivity to prostaglandin E, histamine, hydrocortisone, and X-irradiation with age: studies in a healthy elderly population. *Clin. Immunol. Immunopathol.* 25:243–51, 1982.

114. Shimamura T, Hashimoto K, Sasaki S: Feedback suppression of the immune response in vivo. II. Involvement of prostaglandins in the generation of suppressor-inducer B lymphocytes. *Cell. Immunol.* 69:192–5, 1982.

115. Lukic ML, Cowing C, Leskowitz S: Strain differences in ease of tolerance induction to bovine gamma-globulin: dependence on macrophage function. *J. Immunol.* 114:503–7, 1975.

116. Phipps RP, Scott DW: A novel role for macrophages: the ability of macrophages to tolerize B cells. *J. Immunol.* 131:2122–7, 1983.

117. Diener E, Kraft N, Lee KC, Shiozawa C: Antigen recognition IV: Discrimination by antigen-binding immunocompetent B cells between immunity and tolerance is determined by adherent cells. *J. Exp. Med.* 143:805–21, 1976.

118. Goldings EA: Regulation of B cell tolerance by macrophage-derived mediators: antagonistic effects of prostaglandin E_2 and interleukin 1. *J. Immunol.* 136:817–22, 1986.

119. Herberman RB: *Natural Killer Cell-Mediated Immunity Against Tumors.* Academic Press, N.Y., 1980.

120. Toivanen P, Uksila J, Leino A, Lassila O, Hirvonen T, Ruuskanen O:

Development of mitogen responding T cells and natural killer cells in the human fetus. *Immunol. Rev.* 57:89–105, 1981.

121. Antonelli P, Stewart W, Dupont B: Distribution of natural killer cell activity in peripheral blood, cord blood, thymus, lymph nodes, and spleen, and the effect of in vivo treatment with interferon preparation. *Clin. Immunol. Immunopathol.* 19:161–9, 1981.

122. Timonen T, Ortaldo JR, Herberman RB: Characteristics of human large granular lymphocytes and relationship to natural killer and K cells. *J. Exp. Med.* 153:569–82, 1981.

123. Abo T, Balch CM: A differentiation antigen of human NK and K cells identified by a monoclonal antibody (HNK-1). *J. Immunol.* 124:481–90, 1980.

124. Nieminen P, Paasivuo R, Sakesela E: Effect of a monoclonal antilarge granular lymphocyte antibody on the human NK activity. *J. Immunol.* 128:1097-1101, 1982.

125. Reynolds CW, Timonen T, Herberman RB: Natural killer cell activity in the rat. Isolation and characterization of the effector cells. *J. Immunol.* 127:282–7, 1981.

126. Herberman RB, Ortaldo JR: Natural killer cells: their role in defenses against disease. *Science* 214:24–30, 1981.

127. Bankhurst AD: The modulation of human natural killer cell activity by prostaglandins. *J. Clin. Lab. Immunol.* 7:85–91, 1982.

128. Jondal M, Merrill J, Ullberg M: Monocyte-induced human natural killer cell suppression followed by increased cytotoxic activity during short-term in vitro culture in autologous serum. *Scand. J. Immunol.* 14:555–63, 1981.

129. Droller MJ, Perlmann P, Schneider MU: Enhancement of natural and antibody dependent lymphocyte cytotoxicity by drugs, which inhibit prostaglandin production by tumor target cells. *Cell. Immunol.* 39:154–64, 1978.

130. Droller MJ, Schneider M, Perlmann P: A possible role of prostaglandins in the inhibition of natural and antibody-dependent cell-mediated cytotoxicity against tumor cells. *Cell. Immunol.* 39:165–177, 1978.

131. Merrill JE: Natural killer (NK) and antibody-dependent cellular cytotoxicity (ADCC) activities can be differentiated by their different sensitivities to interferon and prostaglandin E_1. *J. Clin. Immunol.* 3:42–50, 1983.

132. Leung KH, Koren HS: Regulation of human natural killing: II. Protective effect of interferon on NK cells from suppression by PGE_2. *J. Immunol.* 129:1742–7, 1982.

133. Rola-Pleszczynski M, Gagnon L, Sirois P: Leukotriene B_4 augments human natural cytotoxic cell activity. *Biochem. Biophys. Res. Comm.* 113:531–7, 1983.

134. Seaman WE, Woodcock J: Human and murine natural killer cell activity may require lipoxygenation of arachidonic acid. *J. Allergy Clin. Immunol.* 74:407–11, 1984.

135. Bray RA, Brahmi Z: Role of lipoxygenation in human natural killer cell activation. *J. Immunol.* 136:1783–90, 1986.

136. Serhan CN, Hamberg M, Ramstedt, Samuelsson B: Lipoxins: stereochemis-

try, biosynthesis and biological activities. *Adv. Prostagl. Thrombox. Leukotr. Res.* 16:83–97, 1986.

137. Staszak C, Goodwin JS, Troup GM, Pathak DR, Williams RC: Decreased sensitivity to prostaglandin and histamine in lymphocyte from normal HLA-B12 individuals: a possible role in autoimmunity. *J. Immunol.* 125:181–5, 1980.

138. Goodwin JS, Bromberg S, Staszak C, Kaszubowski PA, Messner RP, Neal JF: Effect of physical stress on sensitivity of lymphocytes to inhibition by prostaglandin E_2. *J. Immunol.* 127:518–22, 1981.

139. Grbic JT, Wood JJ, Jordan A, Rodrick ML, Mannick JA: Lymphocytes from burn patients are more sensitive to suppression by prostaglandin E_2. *Surg. Forum* 108–10, 1985.

140. Berenbaum MC, Cope WA, Bundick RV: Synergistic effect of cortisol and prostaglandin E_2 on the PHA response: relation to immunosuppression induced by trauma. *Clin. Exp. Immunol.* 26:534-41, 1976.

141. Rocklin RE, Thistle L, Andera C: Decreased sensitivity of atopic mononuclear cells to prostaglandin E_2 (PGE_2) and prostaglandin D_2 (PGD_2). *J. Immunol.* 135:2033–9, 1985.

5

Inflammation and the neutrophil

Inflammation is a response by which the body repairs tissue damage and defends itself against infection. It is not an immunological response in the traditional sense, since it is nonspecific and can be initiated by completely nonimmune pathways (1–3). Immunological cells and effector mechanisms, however, are commonly participants in the process. The purpose of inflammation is to deliver plasma and cellular components of the blood to extravascular tissues, causing dilution of toxic materials, increase in lymph flow, phagocytosis of damaged tissue and contaminating materials or organisms, and (in the late stages of inflammation) formation of a fibrotic barrier which walls off infection.

Inflammatory responses can result from a variety of immunological processes including a T-lymphocyte-mediated mechanism in allergic contact dermatitis, and a humoral-antibody-mediated process in pemphigus. The cells involved in acute inflammation include mast cells, neutrophils, platelets, and eosinophils, which appear to act in a predictable sequence. The inflammatory sequence is initiated by chemical mediators released from the mast cells, which are found adjacent to small arterioles and in submucosal tissues. These mediators, summarized in Figure 5.1, dilate vessels and initiate a cascade of important events, including participation of the neutrophil in the inflammatory response.

Histamine is the major preformed mast cell mediator, synthesized from the amino acid L-histidine by the action of the enzyme L-histidine decarboxylase. The biological effects of histamine are mediated through two distinct receptor systems, H_1 and H_2. Classical acute vascular inflammatory events are mediated through H_1 receptors, and antiinflammatory effects and vasodilitation are mediated through H_2 receptors (Table 5.1). Thus histamine may activate acute vascular effects, yet inhibit acute cellular inflammation (4). Acute cellular inflammation appears to be mediated by the products of arachidonic acid metabolism.

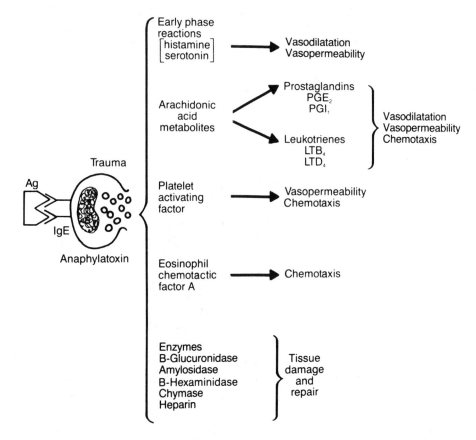

Figure 5.1. Mast cell mediators of inflammation. (Reprinted with permission of the publisher from Sell S: *Basic Immunology: Immune Mechanisms in Health and Disease.* Copyright 1987 by Elsevier, Science Publishing Co., Inc., N.Y., p. 269.)

The exact role of PGs such as PGE_2 in the inflammatory process is difficult to evaluate. For example, in inflammation, the skin becomes red and swollen, and PGE_2 concentrations are increased. Furthermore, the injection of PGE_2 into normal skin produces redness and swelling, and the topical application of cyclooxygenase inhibitors reduces both the redness and swelling of inflammatory lesions. Other data, however, indicate that these symptoms may not be produced directly by PGs. The injection of PGs alone into guinea pig skin (5) or rabbit skin (6) produces very little edema. Williams and Peck studied the effects of intradermal injection of PGs and other mediators on the accumulation of intra-

Table 5.1. A summary of H_1- and H_2-dependent actions of histamine.

H_1 Receptor–Mediated	H_2 Receptor–Mediated	H_1 and H_2
Increased cGMP	Increased cAMP	Vasodilation (hypotension)
Smooth-muscle constriction (bronchi)	Smooth muscle dilation (vascular)	Flush
Increased vascular permeability	Gastric-acid secretion	Headache
Pruritus	Mucous secretion	
Prostaglandin generation	Inhibition of basophil histamine release	
	Inhibition of lymphokine release	
	Inhibition of neutrophil enzyme release	
	Inhibition of eosinophil migration	
	Inhibition of T-lymphocyte–mediated cytotoxicity	
Antagonized by "classical" antihistamines	Antagonized by cimetidine	

Source: Reprinted with permission of the publisher from Sell S: *Basic Immunology: Immune Mechanisms in Health and Disease.* Copyright 1987 by Elsevier Science Publishing Co., Inc., p. 270.

venously injected ^{131}I-albumin in human skin, and concluded that the swelling observed after the injection of PGE_2 must be mediated by some other substance(s), the release or function of which is enhanced in the presence of PGE_2 (7).

It is known that the presence of PGE_2 and $PGF_{2\alpha}$ affect the release and function of histamine. Sondergaard, for example, showed that the whealing response, produced in human skin by PGE_2, can be blocked in part by antihistamines (8). Further work has led to the conclusion that, in general, PGs of the E group potentiate the actions of proinflammatory mediators, while PGs of the F group counteract their effects (1).

Likewise, the LTs and other lipoxygenase products possess potent proinflammatory activities and, like PGE_2 and $PGF_{2\alpha}$, can modulate vascular permeability and activate phagocytes (9). When LTB_4 is injected into rabbit, rat, or human skin, local erythema and extravasation of serum proteins results. This response is greatly augmented by the coadministration of PGE_1, PGE_2, PGD_2, or bradykinin (10–14). Other lipoxygenase products, including 5-, 12-, and 15-HETE produce the same effect; however, they are from 10 to 100 times less active than LTB_4 (10–13).

PROSTAGLANDIN PRODUCTION BY NEUTROPHILS

Each of the cell types involved in chronic and acute cellular inflammation (including platelets, neutrophils, macrophages, and lymphocytes) have been reported to convert arachidonic acid to its active derivatives (1). Neutrophil (PMN) production of these substances appears to be particularly important, however, since PMN are particularly active PG producers and since the activation of these cells precedes the later influx of macrophages and lymphocytes during chronic inflammation.

Neutrophils respond to phagocytic stimuli or to the presence of substances such as chemotactic peptides, in a programmed sequence of secretory events that leads to the elaboration of three classes of mediators (14–16). Probably best studied has been the release of lysosomal hydrolases from neutrophils upon encounter with opsonized antigen. This secretion is normally directed to the interior contents of phagocytic vacuoles for the destruction of the phagocytized substance. When cytochalasin B is utilized to prevent the closure of these phagocytic vacuoles, hydrolases can be recovered outside of the cells. Release of enzymes outside the cell occurs spontaneously in such diseases as rheumatoid arthritis, where the release of neutral proteases (such as collagenase, elastase, and cathepsin G) leads to the inflammatory destruction of connective tissue (14,16,17). Neutrophils also release active products derived from molecular oxygen, such as O_2^-, H_2O_2, OH^-, and O_2, which normally also contribute to the microbicidal system of the phagocytic vacuole (18). These products can also cause inflammation-associated tissue damage. Finally, as a result of the phagocytic process, neutrophils also release stable PGs and LTs (19), which, like the other two classes of mediators, contribute to the inflammatory process.

Weissman has demonstrated that, after ingesting serum-treated zymosan, human neutrophils are stimulated to generate thromboxane B_2 (1). Maximum TXB_2 generation generally occurs before 15 min and in a linear response to zymosan concentration. To test whether phagocytosis was necessary for TXB_2 generation, cytochalasin B was added to inhibit closure of the phagocytic vacuole, thereby preventing phagocytosis. It was found that successful phagocytosis was not necessary for TXB_2 release. Zymosan-stimulated TXB_2 release was inhibited, however, by the addition of indomethacin (1). Indomethacin did not interfere with the release of lysosomal enzymes or the generation of superoxide anions but it was found by this method that serotonin release was linked to the production of TXB_2. As is discussed below, however, the primary product of neutrophil arachidonic acid metabolism is LTB_4.

LEUKOTRIENE SYNTHESIS

The leukotrienes were originally isolated from arachidonic-acid-challenged rabbit PMNs by Borgeat and Samuelsson in 1979 (20). The name leukotriene, in fact, is derived from the name of these leukocytes. Since 1979, it has been shown that neutrophils from a variety of animal species convert arachidonic acid via 5-, 8-,9-,11-, and 12-lipoxygenase pathways to yield a variety of mono- and dihydroxyeicosatetraenoic acids (HETEs) and their corresponding mono- and dihydroxyperoxyeicosatetraenoic acids (HPETEs). All of these lipoxygenase metabolites appear to be biologically active, with the following order of potency: LTs > HPETEs > HETEs. Of all the LTs, LTB_4 has the most potent chemotactic and chemokinetic activities (21).

EFFECT OF PROSTAGLANDINS AND LEUKOTRIENES
ON NEUTROPHILS

As early as 1971, it was demonstrated that the release of lysosomal hydrolases from neutrophils could be inhibited by the presence of stable

Table 5.2. *The effects* in vitro *of leukotriene* B_4
on neutrophil (PMN) activity.

Activity	Reference
Promotes chemotaxis	76,77
Promotes Chemokinesis	30,77
Promotes aggregation	30
Increases adherence	79
Enhances release of lysosomal enzymes	80
Promotes superoxide anion production	81
Stimulates Ca^{2+} influx	78
Elevates intracellular cyclic AMP	79
Enhances C3b receptor expression	80
Enhances complement-dependent cytotoxicity	81

PGs (22,23). PG-treated human cells, for example, do not release lysosomal enzymes when exposed to opsonized particles, chemotactic factors such as N-formyl-L-methionyl-L-leucyl-L-phenylalanine (FMLP) or C5a, con A, or to immune complexes (22,1). This PG-inhibitory effect can also be induced by pretreatment of cells with exogenous cyclic AMP (22). Conversely, agents which elevate intracellular concentrations of cyclic GMP enhance lysosomal/hydrolase release (24). The ability of such agents as PGE_1, PGE_2, cholera enterotoxin, and histamine to inhibit lysosomal enzyme release appears to be directly proportional to their ability to enhance the accumulation of cyclic AMP in neutrophils (24).

In 1971, Kaley and Weiner showed that PGE_1 stimulated the chemotactic migration of rabbit peritoneal granulocytes (25). Subsequent work has suggested that this chemically induced migration is restricted to certain cells only, since peripheral granulocytes from humans, rabbits, or rats, do not respond to PGE_1 (26). TXB_2 has also been reported to be chemotactic for neutrophils (27), as is the cyclooxygenase product 12-L-hydroxy-5,8,10-heptadecatrienoic acid (HHT) (27). These compounds are weak, however, compared to the chemotactic activity of the lipoxygenase products.

The monohydroxy lipoxygenase products, 5-, 12-, and 15-HETE, as well as the HPETEs, are all potent chemotactic and chemokinetic agents for rabbit and human PMN (28). LTB_4, however, is far more potent than any of these (100 times) (29–31). A list of the documented effects of LTB_4 on neutrophils is shown in Table 5.2. When injected intracutaneously in humans, intradermally in rabbits or monkeys, or intraperitoneally in guinea pigs, LTB_4 causes leukocyte accumulation (10,32–

34). It has been suggested that increased PMN adherence to the vascular endothelium (35), followed by chemokinetic and chemotactic activity are involved in these observations. LTB_4 also causes aggregation of PMN *in vitro* (29), an effect that may help explain the neutropenia observed following intravenous injection of this metabolite (35).

The secretion of lysosomal enzymes is increased, in a dose-dependent manner, by nanomolar concentrations of LTB_4 (36,37). This process is energy dependent (38), and may be related to the effect of LTB_4 on PMN calcium flux (39). It is clear, however, from structure–activity studies (40), and from binding experiments (41), that human PMNs have specific membrane receptors for LTB_4. It is thought that other hydroxy-fatty acids, such as 5-HETEs, do not require receptors for their activity and may act, at least in part, through direct covalent insertion into the membrane (42).

A distinct receptor for LTB_4 has been defined on human PMNs using both conventional techniques, and methods utilizing acetone at $-78\ °C$ to extract bound and nonreceptor-bound [^3H]LTB_4 (43). It was found that LTB_4 is bound, stereospecifically, by 26 000–40 000 receptors per cell, with a K_d of 10.8–13.9 nM. Analyses of competitive inhibition of the binding of [^3H]LTB_4 to PMNs by analogs of LTB_4 revealed a close correlation between the chemotactic potency of related 5-lipoxygenase products and their capacity to inhibit the binding of [^3H]LTB_4 (41). Maximally chemotactic concentrations of C5a and N-formylmethionyl peptides, however, do not inhibit PMN binding of [^3H]LTB_4. Nonreceptor binding of [^3H]LTB_4 to PMNs exceeds that of peptide chemotactic factors, and exceeds receptor binding at LTB_4 concentrations higher than 25 nM (41).

LTB_4 is inactivated by granulocytes through ω-oxidation which produces 20-OH LTB_4 and 20-COOH LTB_4 (44). These degradation products appear to have no biological activity (44,45).

LTB_4 functions as a potent inducer of leukocyte infiltration in a variety of models, including neutrophil infiltration into the guinea pig peritoneal cavity (46), granulocyte migration into rabbit skin and hamster cheek pouch (47), granulocyte migration into rabbit eye (48), and granulocyte migration into human skin (13). The topical application of LTB_4 to human skin results in the formation of intraepidermal granulocytic microabscesses, which resemble the lesions observed in pustular psoriasis (49). This observation correlates with the finding that psoriatic lesions contain elevated levels of LTB_4 (50). Elevated LTB_4 levels have been reported in sponge exudates in rats (51), in the gastrointestinal mucosa of patients with inflammatory bowel disease (52), and in the synovial fluid of patients with rheumatoid arthritis or gout (53,54).

REGULATION OF NEUTROPHIL PROSTAGLANDIN AND LEUKOTRIENE PRODUCTION

Many substances must be present for optimum PG synthesis to occur, including appropriate enzymes, oxygen, nonesterified substrate, and hydroperoxide activator (55). Hydroperoxide appears to serve as an initiator of cyclooxygenase activity, and its continued presence is required to achieve maximal enzyme velocity (56). Long-chain fatty acid hydroperoxides are 1000-fold more potent as enzyme stimulators than their nonorganic analog, hydrogen peroxide (57). Phagocytic leukocytes from both humans and guinea pigs produce activator hydroperoxides in quantities large enough to enhance PG synthesis (58). These activators appear to be both H_2O_2 and lipid hydroperoxide. These observations suggest that oxygen metabolites from neutrophils can provide intercellular signals which stimulate PG synthesis by a variety of cell types.

The complement component C5a has been reported to induce PMN degranulation, superoxide anion generation, migration, and cell aggregation via interaction with specific membrane receptors (59,60). Even at concentrations which maximally stimulate these responses, however, C5a induces only a small amount of LTB_4 synthesis, which is now thought to be the main product of neutrophil arachidonic acid metabolism (61,62). Antibody-dependent activation of the complement cascade, with participation of the terminal C5–C9 pathway, on the other hand, induces human PMN to generate significant amounts of LTB_4 (63). This effect seems to be dependent upon C8 (Figure 5.2). Calcium-requirement studies, and measuring calcium uptake and marker flux all indicate that the activation of the cascade leading to LTB_4 synthesis is caused by calcium influx through hydrophilic transmembrane channels (63).

In related studies, it was found that the presence of bacterial lipopolysaccharide reduces both neutrophil LTB_4 receptor, and neutrophil chemotaxis in response to C5a or LTB_4 in rabbits (64). These studies revealed two distinct receptor populations for LTB_4 on normal neutrophils: 10 300 ± 6800 high-affinity binding sites and 85 600 ± 53 000 low-affinity binding sites per cells with mean K_d values of 0.75 ± 0.43 nM, and 70 ± 58 nM respectively (64). Neutrophils from endotoxin-treated rabbits had 68% fewer high-affinity binding sites per cells, with no change in the K_d of either high- or low-affinity receptors, and no change in the number of low-affinity receptors.

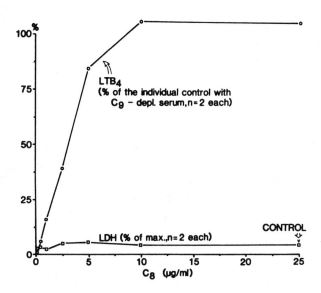

Figure 5.2. Demonstration of the C8-dependence of complement-induced LTB₄ formation. The addition of C8-C9-depleted serum to 6 × 10⁶ antibody-sensitized PMN produced no detectable LTB₄ release (<0.1 resistant ng/10 cells). Reconstitution of C8-C9-depleted serum with C8 restored LTB₄ production, but produced no increase in lactate dehydrogenase (LDH), which is dependent upon the presence of C9. (Reprinted with permission from Seeger W, Suttorp N, Hellwig A, Bhakdi S: Noncytolytic terminal complement complexes may serve as calcium gates to elicit leukotriene B₄ generation in human polymorphonuclear leukocytes. *J. Immunol.* 137:1286–93, 1986.)

Finally, the supernatants of cultures of bacterial-lipopolysaccharide-stimulated human peripheral blood mononuclear cells have been found to prime both eosinophils and neutrophils for subsequent enhanced LT production (Figure 5.3) upon activation with calcium ionophore A23187 (65). The results of this study suggest that a monocyte-derived enhancing factor, produced upon LPS stimulation, is directly effecting arachidonic acid metabolism in the neutrophil.

EFFECT OF NEUTROPHIL PROSTAGLANDINS AND LEUKOTRIENES ON OTHER CELLS

Several groups have investigated the inhibitory effects of neutrophil activity on lymphocyte response (66,67). Especially notable are the studies by Hsu et al., who co-cultured PMNs with autologous lymphocytes and demonstrated that PMN products reduced the PHA responsiveness

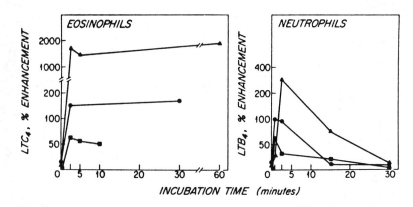

Figure 5.3. Time-dependent enhancing effects of supernatants from LPS-stimulated cultures of peripheral blood mononuclear cells, on the subsequent generation of LTC_4 and LTB_4 by ionophore-activated eosinophils and neutrophils in three experiments. The symbols represent three different granulocyte donors. (Reprinted with permission from Dessein AJ, Lee TH, Elsas P, Ravalese J, Silberstein D, David JR, Austen KF, Lewis RA: Enhancement by monokines of leukotriene generation by human eosinophils and neutrophils stimulated with calcium ionophore A23187. *J. Immunol.* 136:3829–38, 1986.)

of the lymphocytes (66). While these studies did not identify the nature of the inhibitory PMN product, Niwa et al. presented evidence that the suppressive substance may be a PG, since treatment of co-culture PMN with PG synthetase inhibitors reduces their ability to suppress (68). The amount of PGs found in supernatants of untreated PMN/lymphocyte co-cultures was 30 times that obtained by sonification of fresh PMNs, and 10 times higher than found in culture medium containing PMNs alone. Other results suggested a PG-mediated decrease in the number of OKT4(+) lymphocytes in co-cultures, impaired helper-T-cell activity, and enhanced suppressor cell activity, although OKT8(+) cell numbers were not increased (68).

PMN-derived lipoxygenase products likewise seem to affect the behavior of other adjacent unrelated cells. Feinmark and Cannon studied LT synthesis by porcine endothelial cells and found that the intracellular transfer of PMN-produced LTA_4 contributed significantly to endothelial cell LTC_4 production (69). Endothelial cells, in turn, inhibit granulocyte aggregation (70), an activity very important to PMN distribution during inflammation. While this antiaggregation activity of endothelial cells is partially due to prostacyclin (PGI_2) production (Figure 5.4), it is likely

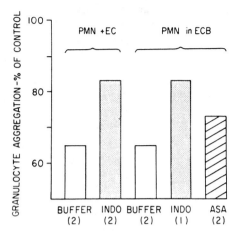

Figure 5.4. Cyclooxygenase inhibition by indomethacin (indo) and aspirin (ASA) partially attenuates the ability of endothelial cells (EC) and endothelial cell culture supernatants (ECB) to modulate granulocyte aggregation. The number of experiments, each using EC monolayers from a different isolate, is shown in parentheses. (Reprinted with permission from Zimmerman GA, Klein-Knoeckel D: Human endothelial cells inhibit granulocyte aggregation in vitro. *J. Immunol.* 136:3839–47, 1986.)

that LT production plays the decisive role (71). Likewise, LTC_4 and LTD_4 have been found to stimulate human endothelial cells to synthesize platelet-activating factor (PAF) and to induce adherence of neutrophils to endothelial cell monolayers (72).

Platelet-activating factor is a phospholipid mediator, synthesized by a variety of cell types including PMN. PAF synthesis has been found to be linked to LTB_4 production by adherent, inflammatory PMN as well as by neutrophils in suspension (73). Results of these studies and others (74,75) support the hypothesis that PAF and LTB_4 are two lipid autocoids derived from a common precursor.

LITERATURE CITED

1. Weissman G: *Prostaglandins and Acute Inflammation*. Current Concepts Series, Upjohn Co, Kalamazoo, 1980.
2. Kuehl FA, Egan RW: Prostaglandins, arachidonic acid, and inflammation. *Science* 210:978–84, 1980.
3. Zurier RB: Prostaglandins and inflammation. *In*: Lee JB (ed.), *Prostaglandins*. Elsevier, N.Y., pp 91–112, 1982.

108   *Prostaglandins, leukotrienes, and the immune response*

4. Sell S: *Basic Immunology: Immune Mechanisms in Health and Disease.* Elsevier, N.Y., pp 261–306, 1987.
5. Horton EW: Action of prostaglandin E_1 on tissues which respond to bradykinin. *Nature* (London) 200:892–3, 1963.
6. Williams TJ: The proinflammatory activity of E-, A-, D- and F-type prostaglandins and analogues 16,16-dimethyl-PGE_2 and (15S)-15-methyl-PGE_2 in rabbit skin: the relationship between potentiation of plasma exudates and local blood flow changes. *Br. J. Pharmacol.* 56:341–2P, 1976.
7. Williams TH, Peck MJ: Role of prostaglandin-mediated vasodilation in inflammation. *Nature* (London) 270:530–2, 1977.
8. Sondergaard J, Greaves MW: Prostaglandin E_1: effect on human cutaneous vasculature and skin histamine. *Br. J. Dermatol.* 84:424–8, 1971.
9. Rola-Pleszczynski M: Immunoregulation by leukotrienes and other lipoxygenase metabolites. *Immunol. Today* 6:302-7, 1985.
10. Higgs GA, Salmon JA, Spayne JA: The inflammatory effects of hydroperoxy and hydroxy acid products of arachidonate lipoxygenase in rabbit skin. *Br. J. Pharmacol.* 74:429–33, 1981.
11. Bray MA, Cunningham FM, Ford-Hutchinson AW, Smith MJH: Leukotriene B_4: a mediator of vascular permeability. *Br. J. Pharmacol.* 72:483–6, 1981.
12. Williams TJ, Jose PJ, Wedmore CV, Peck MJ, Forest MJ: Mechanisms underlying inflammatory edema: the importance of synergism between prostaglandins, leukotrienes, and complement-derived peptides. *Adv. Prostagl. Thrombox. Leukotr. Res.* 11:33–7, 1983.
13. Camp RDR, Coutts AA, Greaves MW, Kay AB, Walport MUJ: Responses on human skin to intradermal injection of leukotrienes. *Br. J. Pharmacol.* 75:168–72P, 1981.
14. Weissmann G: Leukocytes as secretory organs of inflammation. *Hosp. Pract.* 13:53–62, 1978.
15. Weissmann G, Korchak HM, Perez HD, Smolen JE, Goldstein IM, Hoffstein ST: The secretory code of the neutrophil. *J. Reticuloendothel. Soc.* 26:687–700, 1980.
16. Zurier RB, Hoffstein S, Weissmann G: Cytochalasin B: effect on lysosomal enzyme release from human leukocytes. *Proc. Natl. Acad. Sci. USA* 70:844–8, 1973.
17. Weissmann G, Smolen JE, Hoffstein S: Polymorphonuclear leukocytes as secretory organs of inflammation. *J. Invest. Derm.* 71:95–9, 1978.
18. Klebanoff SJ: Antimicrobial systems of the polymorphonuclear leukocytes. *In*: Bellanti JA, Dayton DH (eds.), *The Phagocytic Cell in Host Resistance.* Raven Press, N.Y., pp. 45–59, 1975.
19. Zurier RB, Sayadoff DM: Release of prostaglandins from human polymorphonuclear leukocytes. *Inflammation* 1:93–9, 1975.
20. Borgeat P, Samuelsson B: Transformation of arachidonic acid by rabbit polymorphonuclear leukocytes: formation of a novel dihydroxyeicosatetiaenoic acid. *J. Biol. Chem.* 254:2643–6, 1979.
21. Moore PK: *Prostanoids: Pharmacological, Physiological, and Clinical Relevance.* Cambridge University Press, N.Y., pp. 229–30, 1985.

22. Weissmann G, Dukor P, Zurier RB: Effect of cyclic AMP on release of lysosomal enzymes from phagocytes. *Nature* (London) 231:131–5, 1971.
23. Weissmann G, Zurier RB, Spieler PJ, Goldstein IM: Mechanisms of lysosomal enzyme release from leukocytes exposed to immune complexes and other particles. *J. Exp. Med.* 134:149S, 1971.
24. Zurier RB, Weissmann G, Hoffstein S, Kammerman S, Tai H-H: Mechanisms of lysosomal enzyme release from human leukocytes: II. Effects of cAMP and cGMP, autonomic agonists, and agents which affect microtubule function. *J. Clin. Invest.* 53:297–309, 1974.
25. Kaley G, Weiner R: Effect of prostaglandin E on leukocyte migration. *Nature New Biology* 234:114–15, 1971.
26. Walker JR, Smith MJ, Ford-Hutchinson AW: Prostaglandins and leukotaxis. *J. Pharm. Pharmacol.* 28:745–7, 1976.
27. Goetzl EJ, Gorman RR: Chemotactic and chemokinetic stimulation of human eosinophil and neutrophil polymorphonuclear leukocytes by 12-L-hydroxy-5,8,10-heptadecatrienoic acid (HHT). *J. Immunol.* 120:526–31, 1978.
28. Goetzl EJ, Sun FF: Generation of unique monohydroxyeicosatetraenoic acids from arachidonic acid by human neutrophils. *J. Exp. Med.* 150:406–11, 1979.
29. Ford-Hutchinson AW, Bray MA, Doig MV, Shipley ME, Smith MJ: Leukotriene B: a potent chemokinetic and aggregation substance released from polymorphonuclear leukocytes. *Nature* (London) 286:264–5, 1980.
30. Palmer RMJ, Stepney RJ, Higgs GA, Eakins KE: Chemokinetic activity of arachidonic acid lipoxygenase products on leukocytes of different species. *Prostaglandins* 20:411–18, 1980.
31. Goetzl EJ, Pickett WC: The human PMN leukocyte chemotactic activity of complex hydroxyeicosatetraenoic acids (HETEs). *J. Immunol.* 125:1789–91, 1980.
32. Lewis RA, Goetzl EJ, Drazen JM, Soter NA, Austen KF, Corey EJ: Contractile activities of structural analogs of leukotrienes C and D. Role of the polar substituents. *Proc. Nat. Acad. Sci. USA* 78:4579–83, 1981.
33. Soter NA, Lewis RA, Corey EJ, Austen KF: Local effects of synthetic leukotrienes (LTC$_4$, LTD$_4$, LTE$_4$, and LTB$_4$) in human skin. *J. Invest. Dermatol.* 80:115–19, 1983.
34. Smith MJH, Ford-Hutchinson AW, Bray MA: Leukotriene B$_4$: a potential mediator of inflammation. *J. Pharm. Pharmacol.* 32:517–18, 1980.
35. Bray MA, Ford-Hutchinson AW, Smith MJH: Leukotriene B$_4$: an inflammatory mediator in vivo. *Prostaglandins* 22:213–22, 1981.
36. Hafström I, Palmblad J, Malmsten CL, Radmark O, Samuelsson B: Leukotriene B$_4$ - a stereospecific stimulator for release of lysosomal enzymes from neutrophils. *FEBS Lett.* 130:146–8, 1981.
37. Showell HJ, Naccache PH, Borgeat P, Picard S, Vallerand P, Becker EL, Sha'afi RI: Characterization of the secretory activity of leukotriene B$_4$ toward rabbit neutrophils. *J. Immunol.* 128:811–16, 1982.
38. Bass DA, Thomas MJ, Goetzl EJ, DeChatelet ER, McCall CE: Lipoxygenbase-derived products of arachidonic acid mediate stimulation of hexoseup-

take in human polymorphonuclear leukocytes. *Biochem. Biophys. Res. Commun.* 100:1–7, 1981.

39. Sha'afi RI, Naccache PH, Molski TFP, Borgeat P, Goetzl EJ: Cellular regulatory role of leukotriene B$_4$: its effects on cation homeostasis in rabbit neutrophils. *J. Cell. Physiol.* 108:401–8, 1981.

40. Goetzl EJ, Pickett WC: Novel structural determinants of the human neutrophils chemotactic activity of leukotriene B. *J. Exp. Med.* 153:482–7, 1981.

41. Goldman DW, Goetzl EJ: Specific binding of leukotriene B$_4$ to receptors on human polymorphonuclear leukocytes. *J. Immunol.* 129:1600–4, 1981.

42. Stenson WF, Parker CW: 12-L-Hydroxy-5,8,10,14-eicosatetraenoic acid, a chemotactic fatty acid, is incorporated into neutrophil phospholipids and triglyceride. *Prostaglandins* 18:285–92, 1979.

43. Payan DG, Goldman DW, Goetzl EJ: Biochemical and cellular characteristics of the regulation of human leukocyte function by lipoxygenase products of arachidonic acid. *In*: Chakrin LW, Bailey DM (eds.), *The Leukotrienes.* Academic Press, N.Y., pp 231–45, 1984.

44. Camp RDR, Wollard PM, Mallet AJ, Fincham NJ, Ford-Hutchinson AW, Bray MA: Neutrophil aggregating and chemokinetic properties of 5,12,20-trihydroxy-6,8,10,14-eicosatetraenoic acid isolated from human leukocytes. *Prostaglandins* 23:631–41, 1982.

45. Ford-Hutchinson AW, Rackham A, Zambori R, Rokach J, Ray S: Comparative biological activities of synthetic leukotriene B$_4$ and its omego-oxidation products. *Prostaglandins* 25:29–37, 1983.

46. Smith MJH, Ford-Hutchinson AW, Bray MA: Leukotriene B: a potential mediator of inflammation. *J. Pharm. Pharmacol.* 32:517–18, 1980.

47. Bray MA, Ford-Hutchinson AW, Smith MJH: Leukotriene B$_4$: an inflammatory mediator in vivo. *Prostaglandins* 22:213-22, 1981.

48. Bhattacherjee P, Hammond B, Salmon JA, Stepney R, Eakins KE: Chemotactic response to some arachidonic acid lipoxygenase products in the rabbit eye. *Eur. J. Pharmacol.* 73:21–8, 1981.

49. Camp RDR, Jones RR, Brain S, Wollard P, Greaves M: Production of intraepidermal micro-abscesses by topical application of leukotriene B$_4$: a potential experimental model of psoriasis. *J. Invest. Dermatol.* 82:202–4, 1984.

50. Brain SD, Camp RDR, Dowd RM, Kobza-Black A, Wolland PM, Mallet AI, Greaves MW: Psoriasis and leukotriene B$_4$. *Lancet* ii: 762-3, 1982.

51. Simmons PM, Salmon JA, Moncada S: The release of leukotriene B$_4$ during experimental inflammation. *Biochem. Pharmacol.* 32:1353–9, 1983.

52. Sharon P, Stenson WF: Production of leukotrienes by colonic mucosa from patients with inflammatory bowel disease. *Gastroenterology* 84:1306–13, 1983.

53. Klickstein LB, Shapleigh C, Goetzl EJ: Lipoxygenation of arachidonic acid as a source of polymorphonuclear leukocyte chemotactic factors in synovial fluid and tissue in rheumatoid arthritis and spondyloarthritis. *J. Clin. Invest.* 66:1166–70, 1980.

54. Rae SA, Davidson EM, Smith MJH: Leukotriene B$_4$, an inflammatory mediator in gout. *Lancet* ii: 1122–4, 1982.

55. Kulmacz RJ: Biosynthesis of prostaglandins. In: Willis AL (ed.), *CRC Handbook on Eicosanoids, Prostaglandins and Related Lipids* (in press).
56. Hemler ME, Cook HW, Lands WEM: Prostaglandin biosynthesis can be triggered by lipid peroxides. *Arch. Biochem. Biophys* 193:340–5, 1979.
57. Marshall PJ, Kulmacz RJ, Lands WEM: Hydroperoxides, free radicals, and prostaglandin synthesis. *In:* Bors W, Saran M, Tait D (eds.), *Oxygen Radicals in Chemistry and Biology.* Walter de Gruyter, Berlin, pp 299-304, 1984.
58. Marshall PJ, Lands WEM: In vitro formation of activators for prostaglandin synthesis by neutrophils and macrophages from humans and guinea pigs. *J. Lab. Clin. Med.* 108:525–34, 1986.
59. Chenoweth DE, Goodman MG: The C5a receptor of neutrophils and macrophages. *Agents Actions* 12(suppl):252–273, 1983.
60. Craddock PRD, Hammerschmit D, White JG, Dalmasso AP, Jacob HS: Complement (C5a)-induced granulocyte aggregation in vitro. *J. Clin. Invest.* 60:260–4, 1977.
61. Korchak HM, Vienne K, Rutherford LE, Weissman G: Neutrophil stimulation: receptor, membrane, and metabolic events. *Fed. Proc.* 43:2749–54, 1984.
62. Ward PA, Sulavik MC, Johnson KJ: Activated rat neutrophils. *Am. J. Pathol.* 20:112–20, 1985.
63. Seeger W, Suttorp N, Hellwig A, Bhakdi S: Noncytolytic terminal complement complexes may serve as calcium gates to elicit leukotriene B_4 generation in human polymorphonuclear leukocytes. *J. Immunol.* 137:1286–93, 1986.
64. Goldman DW, Enkel H, Gifford LA, Chenoweth DE, Rosenbaum JT: Lipopolysaccharide modulates receptors for leukotriene B_4, C5a, and formyl-methionyl-leucyl-phenylalanine on rabbit polymorphonuclear leukocytes. *J. Immunol.* 137:1971–6, 1986.
65. Dessein AJ, Lee TH, Elsas P, Ravalese J, Silberstein D, David JR, Austen KF, Lewis RA: Enhancement by monokines of leukotriene generation by human eosinophils and neutrophils stimulated with calcium ionophore A23187. *J. Immunol.* 136:3829–38, 1986.
66. Hsu CCS, Mu-Yen BW, Rivera-Arcilla J: Inhibition of lymphocyte reactivity in vitro by autologous polymorphonuclear cells (PMN). *Cell. Immunol.* 48:288-95, 1979.
67. Jones AL: The effect of polymorphonuclear leukocytes on the blastoid transformation of lymphocytes in mixed leukocyte cultures. *Transplantation* 4:337–43, 1966.
68. Niwa Y, Sakane T, Fukuda Y, Miyachi Y, Kanoh T: Modulation of the immunoreactivity of a T-lymphocyte subpopulation by neutrophil-released prostaglandin. *J. Clin. Lab. Immunol.* 17:37–44, 1985.
69. Feinmark SJ, Cannon PJ: Endothelial cell leukotriene C_4 synthesis results from intercellular transfer of leukotriene A_4 synthesized by polymorphonuclear leukocytes. *J. Biol. Chem.* 261:16466–72, 1986.
70. Zimmerman GA, Klein-Knoeckel D: Human endothelial cells inhibit granulocyte aggregation in vitro. *J. Immunol.* 136:3839–47, 1986.
71. Goetzl EJ, Brindley LL, Goldman DW: Enhancement of human neutrophil

adherence by synthetic leukotriene constituents of slow-reacting substance of anaphylaxis. *Immunology* 50:35–41, 1983.

72. McIntyre TM, Zimmerman GA, Prescott SM: Leukotrienes C_4 and D_4 stimulate human endothelial cells to synthesize platelet-activating factor and bind neutrophils. *Proc. Nat. Acad. Sci. USA* 83:2204–8, 1986.

73. Sisson JH, Prescott SM, McIntyre TM, Zimmerman GA: Production of platelet-activating factor by stimulated human polymorphonuclear leukocytes: correlation of synthesis with release, functional events, and leukotriene B_4 metabolism. *J. Immunol.* 138:3918–26, 1987.

74. Chilton FH, Ellis JM, Olson SC, Wykle RL: 1-0-alkyl-2-arachidonoyl-sn-glycero-3-phosphocholine: a common source of platelet-activating factor and arachidonate in human polymorphonuclear leukocytes. *J. Biol. Chem.* 259:12014–19, 1984.

75. Ramesha CS, Pickett WC: Platelet-activating factor and leukotriene biosynthesis is inhibited in polymorphonuclear leukocytes depleted of arachidonic acid. *J. Biol. Chem.* 261:7592–5, 1986.

76. Rollins TE, Zanolari B, Springer MS, Guindon Y, Zamboni R, Lau CK, Rokach J: Synthetic leukotriene B_4 is a potent chemotaxin but a weak secretagogue for human PMN. *Prostaglandins* 25:281–9, 1983.

77. Serhan CN, Radin A, Smolen JE, Korechak H, Samuelsson B, Weissman G: Leukotriene B_4 is a complete secretagogue in human neutrophils: a kinetic analysis. *Biochem. Biophys. Res. Commun.* 107:1006–12, 1982.

78. Mokki TFP, Naccahe PH, Borgeat P, Sha'afi RI: Similarities in the mechanisms by which formyl-methionyl-leucyl-phenylalamine, arachidonic acid, and leukotriene B_4 increase calcium and sodium influxes in rabbit neutrophils. *Biochem. Biophys. Res. Commun.* 103:227–32, 1981.

79. Claesson HE: Leukotrienes A_4 and B_4 stimulate the formation of cyclic AMP in human leukocytes. *FEBS Lett.* 139:305–8, 1982.

80. Nagy L, Lee TH, Goetzyl EJ, Pickett WC, Kay AB: Complement receptor enhancement and chemotaxis of human neutrophils and eosinophils by leukotrienes and other lipoxygenase products. *Clin. Exp. Immunol.* 47:541–7, 1982.

81. Mogbel R, Sass-Kuhn SP, Goetzl EJ, Kay AB: Enhancement of neutrophil and eosinophil-mediated complement-dependent killing of *Schistosoma mansoni* in vitro by leukotriene B_4. *Clin. Exp. Immunol* 52:519–27, 1983.

6

Malignancy and the arachidonic acid cascade

Arachidonic acid metabolites have been implicated as participants in every stage of the carcinogenic process, and as a probable cause for the lack of an effective antitumor immune response in many cancer patients. The realization that PGs might be involved in the suppression of a tumor-directed immune response had its origin in the work of several groups concerned with the mechanism of cancer-associated hypercalcemia. As early as 1970, experiments by Klein and Raisz demonstrated that, along with parathyroid hormone, PGE_2 was a particularly strong stimulator of bone resorption (1). Using a murine model, Tashjian et al. showed that the secretion of PGE_2 by certain nonmetastatic cancers was associated with severe hypercalcemia (2). The association of PGE_2 production and hypercalcemia has been reconfirmed in humans on many occasions (3). Largely because of this work, it became clear that many naturally occurring and experimentally induced malignant tumors synthesized large quantities of PGs. High concentrations of PGE and/or PGE metabolites have been found in the serum of patients with Hodgkin's and non-Hodgkin's lymphomas (Figure 6.1) (4), breast cancer (5), squamous cell carcinoma of the lung (6), renal cell carcinoma, and a large variety of other malignancies (6,7). In other studies, elevated blood levels of 6-keto-$PGF_{1\alpha}$, a degradation product of PGI_2, have been found to occur in patients with various cancers, but its source has not been identified. An increase in PG levels associated with cancer is generally thought to be as a result of either the increased production of these metabolites directly by the tumor (9) or the increased PG production as part of host monocyte response to the tumor (10). Cancer tissues removed during surgical operations have been reported to contain greater than normal PGE_2 concentrations (11). When cell cultures were established from many tumors, they too, produced large amounts of PGE_2 (12,13). In one study, PGE_2 was demonstrated by means of immunofluorescent staining to occur within human tumors in 27 out of 42 malignancies (14). It is likely that

113

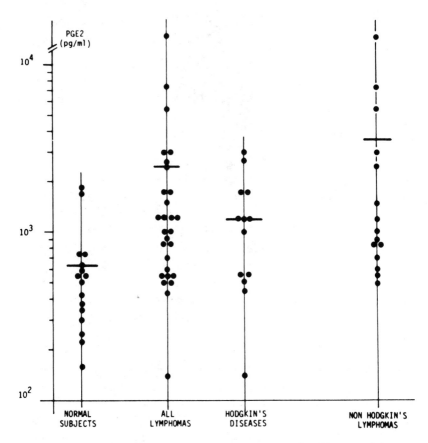

Figure 6.1. Plasma PGE concentrations in normal subjects and in patients with Hodgkin's disease and non-Hodgkin's lymphomas. (Reprinted with permission from Sebahoun G, Maraninchi D, Carcassonne Y: Increased prostaglandin E production in malignant lymphomas. *Acta Haematol.* 74:132–6, 1985.)

the presence of cancer-associated PGs has significance in many aspects of the malignant process including the initiation and promotion of carcinogenesis; the regulation of tumor-cell differentiation, replication, and metastasis; and, probably most significant to the host, the escape of many tumors from elimination by normal immunological surveillance.

THE INITIATION AND PROMOTION OF TUMOR GROWTH

The malignant transformation of cells is a complex process, which can be initiated by a variety of physical, chemical, or biological stimuli. In each case, however, the critical and irreversible step is the change of host DNA

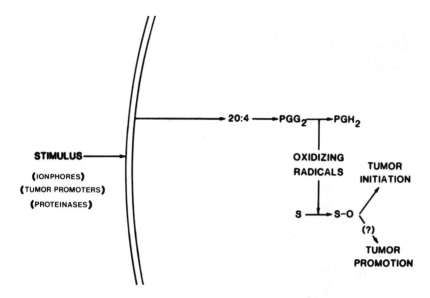

Figure 6.2. Model for prostaglandin-synthetase-dependent cooxygenation of endogenous substrates to produce mutagenic products.
(Reprinted with permission from Marnett LJ, Bienkowski MJ,
Leithauser M, Pagels WR, Panthanickal A, Reed GA: Prostaglandin
synthetase-dependent cooxygenation. *In*: Powles TJ, Bockman RS,
Honn KV, Ramwell P (eds.), *Prostaglandins and Cancer: First International Conference*. Alan R. Liss Inc., New York, pp. 97–111, 1982.)

from normal. This step in the carcinogenic process is termed *initiation*.
Particularly effective in initiating malignant transformation are electrophilic compounds, which can directly modify host nucleic acid. Oxygenation of many of these compounds within the cells can produce
derivatives, which are even more potent than the parent molecule
(15,16). Until recently, it was thought that intracellular oxygenation was
carried out principally by mixed-function oxidases. Marnett et al., however, have shown that potential carcinogens are cooxygenated during the
metabolism of arachidonic acid (17–19). Electrophilic carcinogen formation is thought to occur following the interaction of PGG_2 with microsomal peroxidase (17), which is facilitated by the hydroperoxidase associated with PG synthetase. Such cooxidation has been reported to
occur *in vitro* with a variety of carcinogens including benzo(a)pyrene
(17,18,20), dimethylbenzanthracene and benzanthracene (20,21), benzidine (22), and furantoin compounds (22,23). Recent animal perfusion
studies indicate that this conversion can also occur *in vivo*. Figure 6.2 is a
model of a PG-synthetase-dependent cooxygenation proposed by Mar-

nett et al. (24). A stimulus interacts with the cell plasma membrane, resulting in the release of arachidonic acid synthetase and subsequent PGG_2 biosynthesis. PGG_2 interacts with PG synthetase (or another peroxidase) to release oxidizing radicals, which oxidize endogenous substrates or xenobiotics. The oxidized xenobiotics can be toxic and/or mutagenic. If mutagenic, their production can result in tumor initiation.

Tumor promoters are chemical agents that shorten the latency time in tumor formation. The mechanisms of action of tumor promoters appear to differ from agent to agent and, in general, are poorly understood. It has been suggested, however, that the arachidonic acid cascade may be involved in tumor promotion. This conclusion is based on the observation that the most potent tumor promoters, such as phorbol esters, are extremely potent inducers of PG synthesis (25,26). Chemically related analogs that do not stimulate the arachidonic acid cascade do not promote tumor growth. In addition, the activation of biochemical pathways associated with tumor promotion appears to be PG-dependent. Ornithine decarboxylase (ODC) activity, for example, is depressed by inhibitors of cyclooxygenase and reestablished by the addition of PGE (27–29). In the absence of tumor promoters such as phorbol-esters during this process, PGs have no effect on ODC, suggesting that PGs are modulators of tumor promotion but are not in themselves tumor promoters. Finally, recent evidence also suggests a role for LT production in tumor promotion following phorbol-ester treatment (30) and a general shift of arachidonic acid from the cyclooxygenase pathway to the lipoxygenase pathway.

TUMOR CELL REPLICATION, DIFFERENTIATION, AND METASTASIS

PGs appear to inhibit or stimulate cell replication, depending upon cell type, and the concentration and identity of the metabolite. Following malignant transformation, however, cells appear to be able to escape PG-mediated control of cell replication. The human malignant cell line Lu65, as well as animal tumors such as $HSDM_1$ fibrosarcoma, produce tremendous amounts of PGE_2 and, at the same time, have rapid cell-proliferation rates that are unaffected by the inhibition of PG synthesis (31). The growth of malignant stem cells from patients with chronic myelogenous leukemia are generally refractory to inhibition by PGE_2 in contrast to normal, committed stem cells that are strongly inhibited.

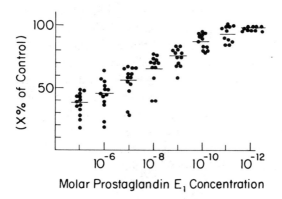

Molar Prostaglandin E₁ Concentration

Figure 6.3. The effects of PGE₁ on day 7 CFU-GM proliferation by 10⁵ nonadherent low-density normal human bone marrow cells. Each point represents the mean of quadruplicate cultures for each normal donor tested. (Reprinted with permission from Pelus L: Antigenic and humoral control of normal and leukemic human myelopoiesis. *In*: Powles TJ, Bockman RS, Honn KV, Ramwell P (eds.), *Prostaglandins and Cancer.* Alan R. Liss Inc., New York, p. 399, 1982.)

The effects of PGs on cell differentiation appear to be opposite to their effects on the cell replication of a specific cell line. For example, while the growth of murine cell line M1 is inhibited by PGE, M1 can be induced to differentiate into mature macrophages by PGE (but not PGF) (32). Similarly, PGE is able to inhibit the expression of Ia antigen in peritoneal macrophages (a differentiation marker), while the growth rate of only Ia-positive macrophages of the (CFU-GM) cell line is inhibited (Figure 6.3) by PGE (33).

The metastatic spread and the growth of cancer distant from the primary tumor is undoubtedly the most important variable in determining the outcome of human malignant disease. Tumor metastasis is dependent upon both host factors and the properties of the tumor cells themselves. To establish a metastasis, tumor cells must invade the tissue surrounding the malignant site, to the point where they eventually reach the blood vessels or lymphatic circulation. Once in either circulatory system, malignant cells must survive host defenses to reach a distant capillary bed and then again proliferate to form a new tumor. Interruption of any of these sequential steps can inhibit metastasis.

It now appears that arachidonic acid metabolites are involved in the metastatic process, at least in an indirect way, through control of platelet aggregation. Sloan et al. found that some metastatic tumors release cysteine protease (cathepsin B), an enzyme that induces platelet thromboxane (TX) synthesis, suggesting a direct influence of malignant cells on platelets (34). Honn et al. found that PGI_2 inhibited the metastasis of Lewis lung carcinoma and B16 melanoma in inbred mice (35,36). The addition of a PGI_2 enhancing agent, Nafazatrom, increased the antimetastatic effects. Furthermore, tumor cells were found to cause the aggregation of human platelets with release of TXA_2 (37). Honn et al. hypothesized that the intravascular balance between PGI_2 and TXA_2 is disrupted in favor of platelet aggregation as part of tumor metastasis, and that selective inhibition of TX synthesis should reduce tumor metastasis (38).

That platelet aggregation is associated with metastasis is also suggested by experiments with several distinct cell lines of the same malignancy. It has been observed that the extremely metastatic $B16F_{10}$ melanoma cell line produces less PGD_2 than the moderately metastatic B16F parent cell line (39). These results are summarized in Figure 6.4. Though not as active as PGI_2, PGD_2 also inhibits platelet aggregation, giving further circumstantial evidence in support of a connection between platelet aggregation and metastasis. Finally, another study using four variants of a murine sarcoma showed that the two highly metastatic sublines released significantly less 6-keto-$PGF_{1\alpha}$ and more TXB_2 than two poorly metastatic sublines (40). As with the other examples cited, the two metastatic variants induced platelet aggregation *in vitro* as a result of TX synthesis.

PROSTAGLANDINS/LEUKOTRIENES AND IMMUNE
SURVEILLANCE

The question of resistance to malignancy by virtue of host immunological surveillance has been debated since the modification and elaboration of this theory in 1970 by Burnet (41). Burnet described immune surveillance as

> the concept that a major function of the immunological mechanisms in mammals is to recognize and eliminate foreign patterns arising in (the) body by somatic mutation or by some equivalent process. It will be evident that the thymus-dependent system of immunocytes will be almost solely responsible for surveillance, antibody and antibody-producing cells having an almost negligible role (41).

It now seems clear that immune surveillance is indeed a valid principle and is particularly important in the case of virus-induced tumors. The

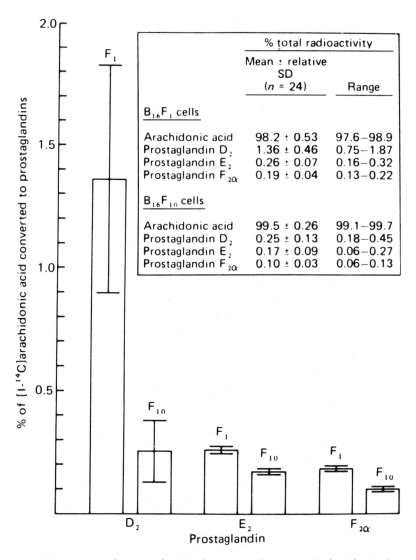

	% total radioactivity	
	Mean ± relative SD (n = 24)	Range
$B_{16}F_1$ cells		
Arachidonic acid	98.2 ± 0.53	97.6–98.9
Prostaglandin D_2	1.36 ± 0.46	0.75–1.87
Prostaglandin E_2	0.26 ± 0.07	0.16–0.32
Prostaglandin $F_{2\alpha}$	0.19 ± 0.04	0.13–0.22
$B_{16}F_{10}$ cells		
Arachidonic acid	99.5 ± 0.26	99.1–99.7
Prostaglandin D_2	0.25 ± 0.13	0.18–0.45
Prostaglandin E_2	0.17 ± 0.09	0.06–0.27
Prostaglandin $F_{2\alpha}$	0.10 ± 0.03	0.06–0.13

Figure 6.4. Production of PGD_2 by B16 melanomas. Under identical conditions, B16F1 cells produced significantly more ($p < 0.001$) PGD_2 than did B16F10 cells. (Reprinted with permission from Fitzpatrick FA, Stringfellow DA: Prostaglandin D_2 formation by malignant melanoma cells correlates inversely with cellular metastatic potential. *Proc. Natl. Acad. Sci. USA* 76:1765–9, 1979.)

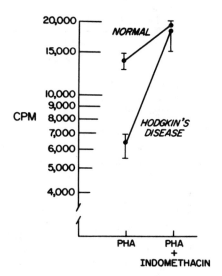

Figure 6.5. Phytohemagglutinin (PHA) response in counts per minute (CPM) of 6 patients with Hodgkin's disease and 20 normal controls with or without the addition of indomethacin. Without indomethacin, the patient and control group are significantly different ($t = 3.01$, $p < 0.002$). After indomethacin there is no significant difference ($t = 0.33$, $p > 0.6$). (Reprinted with permission from Goodwin JS, Messner RP, Bankhurst AD, Peake GT, Saiki JH, Williams RC: Prostaglandin producing suppressor cells in Hodgkin's disease. *N. Engl. J. Med.* 297:963–8, 1977.)

definition, therefore, of the mechanisms by which host immune response, and particularly T-lymphocyte response, is altered in favor of the tumor has important clinical implications.

The profound depression of T-lymphocyte response in cancer patients and in tumor-bearing animals is frequently observed. Pelus and Strausser discovered that the depressed mitogen response of splenic lymphocytes taken from tumor-bearing mice could be restored by the addition of indomethacin to the cultures (42). The work of Twomey et al. (43) and Sibbitt et al. (44) suggested that activation of suppressor cells was responsible for a similar depressed lymphocyte response in patients with Hodgkin's disease. Studying the suppression of the PHA response in patients with Hodgkin's disease, Goodwin et al. found that the addition of indomethacin, or the PG-synthetase inhibitor RO-20-5720, produced a statistically significant improvement in [3H]thymidine incorporation

(Figure 6.5) to almost normal levels (45). This suppression of the PHA response could also be reversed by the removal of glass-adherent cells prior to culture; this is not surprising since PG-producing suppressor cells are glass adherent (46). Goodwin demonstrated, and Bockman later confirmed, that the suppression of the lymphocyte response in patients with Hodgkin's disease was due to increased PGE_2 production by the glass-adherent mononuclear suppressor cells, rather than to an increase in lymphocyte susceptibility to the effects of PG (47). Production of PGE_2 and concomitant loss of T-lymphocyte colony formation appear to be related to the severity of the cancer; a progressive reduction in lymphocyte colony count is observed with advancing stage of the disease. In cases of advanced disease, adherent cells from patients with Hodgkin's disease produced PGE_2 at approximately four times the normal rate (47).

While the evidence is quite convincing that increased monocyte production of PGE_2 in Hodgkin's disease is a major cause of T-cell dysfunction, a similar association in other cancers is sometimes less clear. However, as Figure 6.6 indicates, experiments have shown that several types of murine tumors induce increased PGE production by host macrophages (48). Peripheral blood monocytes from human patients with cancer of the head and neck, produce significantly more PGE_2 than monocytes from normal individuals, and it has been found that this increase correlates with decreased lymphoproliferation *in vitro* (10). Increased PGE production was also observed in monocytes from patients with non-Hodgkin's lymphoma (4) and colonic carcinoma (49) and excessive monocyte suppression has been reported in patients with lung or breast cancer (50), bladder or prostate cancer (51), and malignant melanoma (52).

In addition to inducing monocyte/macrophage production of PGE_2, many types of tumors produce large amounts of PG themselves, and it is probable that this is also a mechanism by which they escape from immune surveillance (9,53). This was clearly shown to be the case by Plescia et al. (54) in their work with two murine tumors, MCDV-12 (a virus-induced ascites cell line) and MC-16 (a methylcholanthrene-induced fibrosarcoma cell line). Mice bearing either of these tumors were found to be unresponsive to immunization with sheep erythrocytes. That the tumor cells were directly responsible for this unresponsiveness was demonstrated by experiments *in vitro* in which the generation of plaque-forming cells was reduced by the addition of MCDV-12 or MC-16 tumor cells to splenic lymphocytes stimulated with sheep erythrocytes (54).

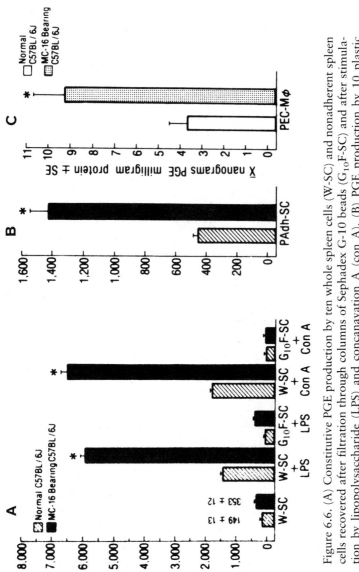

Figure 6.6. (A) Constitutive PGE production by ten whole spleen cells (W-SC) and nonadherent spleen cells recovered after filtration through columns of Sephadex G-10 beads ($G_{10}F$-SC) and after stimulation by lipopolysaccharide (LPS) and concanavation A (con A). (B) PGE production by 10 plastic adherent spleen cells (PAdh-SC) from normal and tumor-bearing mice. (C) PGE production by 2×10 adherent peritoneal macrophage cells (PEC-M) isolated from control and MC-16-bearing mice. (Reprinted with permission from Pelus C, Bockman R: Increased prostaglandin synthesis by macrophages from tumor-bearing mice. *J. Immunol.* 123:2118–25, 1979.)

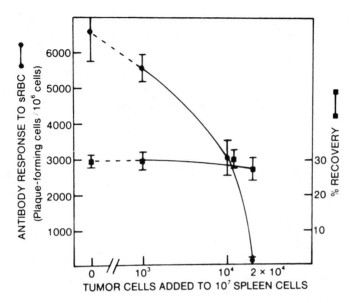

Figure 6.7. Immunosuppressive property of syngeneic methylcholan-threne-induced tumor cells (MC 16). Tumor cells were added to cultures *in vitro* of syngeneic C^{57}B1/6J spleen cells and SRBC. After 4 days, the cultures were examined for plaque-forming cells and viable spleen cells. (Reprinted with permission from Plescia OJ, Smith AH, Grinwich K: Subversion of the immune system by tumor cells and the role of prostaglandins. *Proc. Natl. Acad. Sci. USA* 72:1848–53, 1975.)

Figure 6.7 shows the results obtained after 4 days when cultures were examined for plaque-forming cells and viable spleen cells. Both MCDV-12 and MC-16 tumor cell lines secrete large amounts of PGE$_2$ (55), and it was found that the addition of submicromolar concentrations of PGE$_2$ instead of tumor cells to the plaque assay resulted in the same suppressive effect (54). Tumor-induced immunosuppression could be partially blocked *in vitro* and *in vivo* with indomethacin (54) and other PG-synthetase inhibitors (55). Finally, Plescia et al. discovered that tumor growth *in vivo* could be retarded in the mice with daily injections of indomethacin, presumably because immune reactivity was restored (54,55).

Encouragingly, Plescia's observations have their human counterpart. Droller et al. demonstrated that PG-producing human bladder tumor cell lines were capable of inhibiting host antibody-dependent cytotoxicity (ADCC) and natural killer (NK) cell activity, and that the administration of PG-synthetase inhibitors resulted in an increase in both types of cytotoxicity against the tumor target cells (56). Likewise, Brunda et al. showed that splenic lymphocytes from mice with tumors induced by the Maloney sarcoma virus developed severely depressed NK activity; cytotoxicity could be restored to normal by the administration of aspirin or indomethacin (57). These results are shown in Tables 6.1 and 6.2.

There are data, however, which discourage the hope that PG-synthetase inhibitors could be used as effective antitumor agents. By way of explanation, it is important to remember that of all the products of cyclooxygenase activity PGE appears to be especially important because of its immunoregulatory potential. Suppression of PGE synthesis would be expected to have a positive effect on the immunological mechanisms whereby tumors are recognized and eliminated. PGE may, however, also have a direct antitumor growth effect, at least for those tumors which have specific PGE receptors. Exposure of such cells to PGE would increase the intracellular cyclic AMP levels, and thereby down regulate growth. In this case, inhibition of PGE synthesis would promote tumor growth (58). Other products of arachidonic acid metabolism, also have an effect on the homeostatic balance between tumor and host. Production of PGD, for example, is important in controlling cell proliferation. Suppression of PGD synthesis would be expected to increase rather than decrease tumor growth. It is no surprise, therefore, that tumor treatment with cyclooxygenase inhibitors has had mixed success. Seemingly contradictory reports can sometimes be explained on the basis of the model selected. For example, immune response to the murine B-16 melanoma line, which produces primarily PGD_2 instead of PGE, has been reported to be unaffected or even further suppressed by indomethacin treatment (Figure 6.8) and the growth of the tumor accelerated (59). Stringfellow and Fitzpatrick also showed that PGD_2 production by the tumor cells was inversely proportional to their metastatic potential (60). Exposure of the B-16 melanoma cells to indomethacin prior to injection resulted in an increase in pulmonary metastasis. PGD_2 seemed to have a direct antimetastatic effect on mouse melanoma (60), and an antiproliferative effect on melanoma cells *in vitro* (61). PGD_2 also can inhibit the proliferation of human neuroblastoma cells (62) and of leukemia cells *in vitro* (63). Experiments such as these have led to the suggestion that PGD_2 be administered as an antineoplastic agent (63).

Finally, to further confuse an already confused area, Lynch and Salomon were able to confirm that the administration of PG-synthetase inhibitors slowed the growth of many experimental tumors; however, they did not feel that the drugs' modus operandi was via stimulation of

Table 6.1. *NK cytotoxicity of spleen cells from tumor bearing mice: effect of indomethacin.*

Treatment	CBA		Swiss nu/nu
	Expt. 1	Expt. 2	Expt. 3
None	28.8 ± 0.8	12.0 ± 1.2	34.0 ± 1.0
Indomethacin	N.T.	9.2 ± 0.2	33.8 ± 0.4
MSV	13.5 ± 0.5	7.0 ± 0.4	17.4 ± 0.9
MSV and indo-methacin	23.0 ± 0.8	11.5 ± 0.6	26.1 ± 0.8
MSV and EtOH	17.9 ± 0.5	5.0 ± 0.6	N.T.

Note: MSV, Maloney sarcoma virus; EtOH, ethanol (the diluent for indomethacin).

Source: Reprinted with permission from Brunda MJ, Herberman RB, Holden HT: Inhibition of natural killer cell activity by prostaglandins. *J. Immunol.* 124:2682–7, 1980.

Table 6.2. *NK cytotoxicity of spleen cells from tumor-bearing mice: effect of aspirin.*

Treatment	CBA	nu/nu	
	Expt. 1	Expt. 2	Expt. 3
None	41.0 ± 1.7	31.7 ± 1.4	55.4 ± 1.3
Aspirin	36.0 ± 3.6	35.1 ± 0.6	47.0 ± 0.8
MSV	15.6 ± 0.6	16.7 ± 0.4	31.7 ± 1.4
MSV, BSS	N.T.	14.1 ± 0.2	39.0 ± 0.4
MSV, Aspirin	32.6 ± 1.0	33.1 ± 1.1	52.3 ± 1.2

Note: MSV, Maloney sarcoma virus; BSS, balanced salt solution.

Source: Reprinted with permission from Brunda MJ, Herberman RB, Holden HT: Inhibition of natural killer cell activity by prostaglandins. *J. Immunol.* 124:2682–7, 1980.

the immune response (64), as mice that had completely eliminated their tumors through indomethacin treatment were not resistant to rechallenge. Furthermore, augmentation of the immune response by indomethacin did not correlate with the antitumor effects of the drug. Goodwin, however, feels that these objections are not significant as the data can be explained by taking into consideration the stimulatory effect of the PG-synthestase inhibitors on NK activity and macrophage cytotoxicity, neither of which demonstrates immunological memory (65). As Goodwin also points out, the only feature common to all of the drugs in these studies that have shown antitumor activity (indomethacin, aspirin,

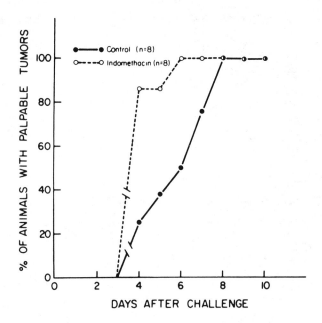

Figure 6.8. The effect of indomethacin (100 ug/day intraperitoneally) on the rate of appearance of subcutaneous B-16 melanomas. Mice were inoculated with 5×10^5 viable cells. The indomethacin curve is significantly different from the control ($p < 0.05$). (Reprinted with permission from Favalli C, Garaci E, Etheredge E, Santoro MG, Jaffe B: Influence of PGE on the immune response in melanoma bearing mice. *J. Immunol.* 125:897–902, 1980.)

RO-20-5720, flurbiprofen, xylenol, etc.), is the ability to inhibit PG production (65).

Finally, evidence is now accumulating that lipoxygenation of arachidonic acid may be required for antitumor cytotoxic responses to occur (66,67). Clearly, lipoxygenase products are able to modulate NK activity *in vitro* (67,68). For example, recent studies have shown that nordihydroguaiaretic, quercetin, eicosatetraynoic acid, phenidone, and esculetin (agents known to inhibit cellular lipoxygenase activity) also inhibit human natural killer cell-mediated cytotoxicity of K562 tumor target cells in a dose-dependent fashion (68). Of the products of lipoxygenation, it appears that 5-HPETE and LTB are the most effective enhancers of NK activity. Conversely, PG production in general, and PGE_2 production specifically, seem to inhibit NK activity. Roder and Klein showed that the addition of indomethacin could restore the NK-mediated cytolysis of YAC tumor target cells (69). Brunda et al. confirmed these observations and extended them to show that PGE inhibited and indomethacin enhanced interferon-augmented NK activity against tumor target cells (57). Conditions favoring lipoxygenase activity rather than cyclooxygenase activity in the tumor-bearing host would, therefore, logically favor tumor homeostasis or elimination via the NK cytotoxic response. Likewise, direct macrophage tumoricidal activity may be important in tumor immunity. Schulz et al. showed that exogenously added PGE_2 inhibits interferon-induced macrophage tumor cell killing (70); Murray reported that human monocyte, but not lymphocyte, ADCC could be augmented by inhibitors of PGE_2 synthesis (71). There is no general agreement, however, concerning the role of each of these forms of cytotoxicity in resistance to specific malignancies.

DIET AND RESISTANCE TO CANCER

It is now thought that the risk of developing certain types of cancer is related to diet. Excellent reviews that have dealt with this subject in detail are available (72,73). Recent reports have made this connection for breast, colon, prostate, pancreatic, endometrial, and ovarian cancer. It appears that cancer rates are higher in people whose diets are high in fat and low in fruits, vegetables, whole grains, and fiber-rich foods. It has also been reported that the composition and type of fat consumed may have as much influence on the development of cancer as the amount of fat consumed. Fats that contain polyunsaturated fatty acids of the omega-6 family seem to favor the growth of tumor cells. Animal experiments have shown that linoleic acid seems to be an especially important member of the omega-6 family, and that the high consumption of this fatty acid encourages the malignant process (74–76). While the mechanisms of fatty acid enhancement of tumor growth are not completely clear, it is known that polyunsaturated fatty acids can readily oxidize to form a variety of potential mutagens, tumor promoters, and carcinogens (74,77). Among these are fatty acid hydroperoxides, endoperoxides, enals, aldehydes, and alkoxy- and hydroperoxyradicals, which promote the growth of cancer cells. Polyunsaturated fatty acids, such as linoleic acid, also act as precursors for arachidonic acid and its metabolites, and, in fact, there is evidence that omega-6 fatty acids promote cancer mainly through their ability to elicit the production of immunosuppressive PGs such as PGE_2 (74,76). Unfortunately, it is impossible to eliminate omega-6 polyunsaturated fatty acids from the human diet because they are needed for normal biochemical functions. There is even occasional encouragement to increase the intake of these compounds since they tend to reduce serum cholesterol levels and reduce coronary heart disease (78,79).

Epidemiological studies of Greenland Eskimos, the Japanese, and Icelanders have suggested that the consumption of large amounts of seafood reduces the incidence of coronary heart disease, atherosclerosis, hypertension, and some types of cancer (such as that of the breast and colon) (75,76). Changing the dietary habits of individuals from these populations results in an increase in cancer incidence and mortality.

Fish oils from deep-dwelling, cold-water fish such as mackerel, blue fish, and herring contain low levels of omega-6 fatty acids and high levels of omega-3 fatty acids such as eicosapentaenoic acid and docosahexacenoic acid (Figure 6.9). Work with animal models has now shown that supplementing diets directly with these omega-3 fatty acids can retard the growth of tumor cells. Fish-oil-enriched diets decrease the formation of PGE_2, an effect which corresponds to the tumor resistance (76). Macrophages have been shown to take up dietary omega-3 fatty acids, thus reducing cellular arachidonic acid levels, their ability to produce PGE_2, and their immunoregulatory activity. In addition, it appears that omega-3 fatty acids also enhance the macrophage production of arginase, an enzyme that has cytolytic activity against tumor cells. Finally, omega-3 polyunsaturated fatty acids may, by inhibiting cyclooxygenase and reducing PGE_2 synthesis, divert arachidonic acid into the lipoxygenase pathway to produce compounds such as 5-HETE and the LTs.

It has been suggested that the LTs function as chemotactic agents for macrophages and other immunological cells capable of controlling tumor growth, and it has been shown that 5-HETE inhibits the growth of tumor cells (74,80). Another nutritional approach to the potential tumor-promoting influence of the peroxidation of arachidonic acid is the administration of biological antioxidants as part of the diet. Dietary vitamin E and selenium have been reported to protect cell membranes from peroxidative damage (81,82) and also appear to inhibit PG production directly. In addition, vitamin C is a biological redox reagent that interferes with the oxidative process, particularly during the inflammatory response, and may have a role in modifying arachidonic acid metabolism as well (83). The increased intake of each of these compounds has been advocated as an adjunct to preventing or retarding the malignant process (84). One would expect their effect to be similar to the effect produced by indomethacin in the various tumor models.

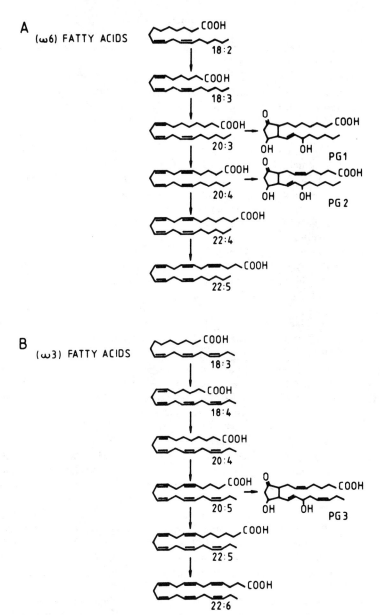

Figure 6.9. Omega-6 fatty acids and their products, the monoenoic and dienoic PGs. Omega-3 fatty acids and their products, the trienoic PGs. (Reprinted with permission from Koivistoinen P, Hyvonen L: Effect of dietary factors on prostanoids. *Ann. Clin Res.* 16:234–40, 1984.)

LITERATURE CITED

1. Klein DC, Raisz LG: Prostaglandins: stimulation of bone resorption in tissue culture. *Endocrinology* 86:1436–70, 1970.
2. Tashjian AH, Voelkel EF, Levine L, Goldhaber PG: Evidence that bone resorption-stimulating factor produced by mouse fibrosarcoma cells is prostaglandin E_2. *J. Exp. Med.* 136:1329–43, 1972.
3. Seyberth HW, Raisz LG, Oates JA: Prostaglandins and hypercalcemia states. *Ann. Rev. Med.* 29:23–9, 1978.
4. Sebahoun G, Maraninchi D, Carcassonne Y: Increased prostaglandin E production in malignant lymphomas. *Acta Haemat.* 74:132–6, 1985.
5. Powles T, Coomes RC, Neville AM, Ford HT, Gazet JG, Levine L: 15-keto-13,14-dihydroprostaglandin E_2 concentrations in serum of patients with breast cancer. *Lancet* ii:138, 1977.
6. Seyberth HW, Segre GW, Morgan JL, Sweetman BJ, Potts JT, Oates JA: Prostaglandins as mediators of hypercalcemia associated with certain types of cancer. *N. Engl. J. Med.* 293:1278–83, 1975.
7. Demers LM, Allegra JC, Harvey HA, Lipton A, Luderer JR, Mortel R, Brener PE: Plasma prostaglandins in hypercalcemic patients with neoplastic disease. *Cancer* 39:1559–62, 1977.
8. Bennett A: Prostanoids and cancer. *Ann. Clin. Res.* 16:314-17, 1984.
9. Burchiel SW, Rubin M, Giorgi J, Peake G, Warner H: Prostaglandin production by murine tumors. *In*: Wolfe P (ed.), *The Importance of Tumor Markers in Clinical Medicine*, Masson Corp., New York, pp. 133–44, 1979.
10. Berlinger NT: Deficient immunity in head and neck cancer due to excessive monocyte production of prostaglandins. *Laryngoscope* 94:1407–10, 1984.
11. Bennett A, Del Tacca M, Stanford IF, Zebro T: Prostaglandins from tumors of human large bowel. *Br. J. Cancer* 35:881–5, 1977.
12. Bennett A, McDonald AM, Simpson JS, Stamford IF: Breast cancer, prostaglandins, and bone metastases. *Lancet* i:218–20, 1975.
13. Rolland PH, Martin PM, Jacquemier J, Rolland AM, Toga M: Prostaglandins in human breast cancer: evidence suggesting that an elevated prostaglandin production is a marker of high metastatic potential for neoplastic cells. *J. Natl. Cancer Inst.* 64:1061–70, 1980.
14. Husby G, Strickland RG, Rigler GL, Peake GT, Williams RC Jr: Direct immunochemical detection of prostaglandin and cyclic nucleotides in human malignant tumors. *Cancer* 40:1629–40, 1977.
15. Miller EC: Studies on the formation of protein-bound derivatives of 3:4-benzypyrene in epidermal fraction of mouse skin. *Cancer Res.* 11:100–8, 1951.
16. Heidelberger C, Weiss SM: The distribution of radioactivity in mice following administration of 3:4-benzypyrene-5-/ul4/dc and 1:2,5:6-dibenzanthracene-9,10-14C. *Cancer Res.* 11:885–91, 1951.
17. Marnett LJ, Wlodaver P, Samuelsson B: Cooxygenation of organic substrates by the prostaglandin synthetase of sheep vesicular gland. *J. Biol. Chem.* 250:8510–17, 1975.

18. Marnett LJ, Reed GA, Johnson JT: Prostaglandin synthetase dependent benzo(a)pyrene oxidation: products of the oxidation and inhibition of their formation by antioxidants. *Biochem. Biophys. Res. Commun.* 79:569–76, 1977.
19. Zenser TV, Mattammal MB, Davis BB: Mechanism of FANFT cooxydation by prostaglandin cyclooxygenase from rabbit inner medulla. *Kidney International.* 16:688–94, 1979.
20. Sivarajah K, Anderson MW, Eling TE: Metabolism of benzo(a)pyrene to reactive intermediates via prostaglandin biosynthesis. *Life Sci.* 23:2571–8, 1978.
21. Sivarajah K, Lasker JM, Eling TE: Prostaglandin synthetase dependent co-oxydation of (±)-benzo(a)pyrene-7,8-dihydrodiol by human lung and other mammalian tissues. *Cancer Res.* 43:2632–6, 1983.
22. Zenser TV, Mattammal MB, Armbrecht HJ, Davis BB: Benzidine binding to nucleic acids mediated by the peroxidative activity of prostaglandin endoperoxide synthetase. *Cancer Res.* 40:2839–45, 1980.
23. Zenser TV, Mattammal MB, Davis BB: Co-oxidation of benzidine by renal medullary prostaglandin cyclooxygenase. *J. Pharmacol. Exp. Ther.* 211:460–4, 1980.
24. Marnett LJ, Bienkowski MJ, Leithauser M, Pagels WR, Panthanickal A, Reed GA: Prostaglandin synthetase-dependent cooxygenation. *In:* Powles TJ, Bockman RS, Honn KV, Ramwell P (eds.), *Prostaglandins and Cancer: First International Conference.* Alan R. Liss Inc, New York pp. 97–111, 1982.
25. Levine L, Hassid A: Effects of phorbol-12,13-diesters on prostaglandin production and phospholipase activity in canine kidney (MDCK) cells. *Biochem. Biophys. Res. Commun.* 79:477–84, 1977.
26. Ashendel CL, Boutwell RK: Prostaglandin E and F levels in mouse epidermis are increased by tumor-promoting phorbol esters. *Biochem. Biophys. Res. Commun.* 90:623–7, 1979.
27. Verma AK, Rice HM, Boutwell RK: Prostaglandins and skin tumor promotion: inhibition of tumor promoter induced ornithine decarboxylase activity by inhibitors of prostaglandin synthesis. *Biochem. Biophys. Res. Commun.* 79:1160–6, 1977.
28. Verma AK, Ashendel CL, Boutwell RK: Inhibition by prostaglandin synthesis inhibitors of the induction of epidermal ornithine decarboxylase activity, the accumulation of prostaglandins, and tumor promotion caused by 12-0-tetra-decanoylphorbol-13-acetate. *Cancer Res.* 40:308–15, 1980.
29. Furstenberger G, Marks F: Indomethacin inhibition of cell proliferation induced by the phorbol ester TPA is reversed by prostaglandin E_2 in mouse epidermis in vivo. *Biochem. Biophys. Res. Commun.* 84:1103–11, 1978.
30. Fischer SM, Gleason GL, Mills GD, Slaga TL: Indomethacin enhancement of TPA tumor promotion in mice. *Cancer Lett.* 10:343–50, 1980.
31. Bockman R, Bellin A, Repo MA, Hickok N, Kameya T: In vivo and in vitro biological activities of two human cell lines derived from anaplastic lung cancers. *Cancer Res.* 43:4511–16, 1983.

32. Ichikawa Y: Differentiation of a cell line in myeloid leukemia. *J. Cell Physiol.* 74:233–4, 1969.
33. Pelus L: Antigenic and humoral control of normal and leukemic human myelopoiesis. *In*: Powles TJ, Bockman RS, Honn KV, Ramwell P (eds.), *Prostaglandins and Cancer.* Alan R. Liss Inc., New York, pp. 399–413, 1982.
34. Sloane B, Dunn J, Honn KV: Lysosomal cathepsin B: a correlation with metastatic potential. *Science* 212:1151–3, 1981.
35. Honn KV, Cicone B, Skoff A: Prostacyclin: a potent antimetastatic agent. *Science* 212:1270–2, 1981.
36. Honn KV, Cavanaugh P, Evens C, Taylor JD, Sloane BF: Tumor cell-platelet aggregation: induced by cathepsin B-line proteinase and inhibited by prostacyclin. *Science* 217:540–2, 1982.
37. Menter D, Neagos G, Dunn J, Palazzo R, Tchen TT, Taylor JD, Honn KV: Tumor cell-induced platelet aggregation: inhibition by prostacyclin, thromboxane A$_2$, and phosphodiesterase inhibitors. *In*: Powles TJ, Bockman RS, Honn KV, Ramwell P (eds.), *Prostaglandins and Cancer: First International Conference.* Alan R. Liss Inc. New York, pp. 809–13, 1982.
38. Honn KV, Buese WD, Sloane BF: Prostacyclin and thromboxanes: implications for their role in tumor cell metastasis. *Biochem. Pharmacol.* 32:1–11, 1983.
39. Fitzpatrick FA, Stringfellow DA: Prostaglandin D$_2$ formation by malignant melanoma cells correlates inversely with cellular metastatic potential. *Proc. Natl. Acad. Sci. USA* 76:1765–9, 1979.
40. Donati MB, Borowska A, Bottazzi B: Metastatic potential correlates with changes in the thromboxane-prostacyclin balance. *In*: Samuelsson B, Paoletti R, Ramwell PW (eds.), *Fifth International Conference on Prostaglandins* (Florence, Italy, 1982), Raven Press, New York, 1983.
41. Burnet FM: *Immunological Surveillance.* Pergamon Press, Oxford, 1970.
42. Pelus LM, Stausser HR: Indomethacin enhancement of spleen-cell responsiveness to mitogen stimulation in tumorous mice. *Int. J. Cancer* 18:653-60 , 1976.
43. Twomey JJ, Laughter AH, Farrow S, Douglass CC: Hodgkin's disease: an immunodepleting and immunosuppressive disorder. *J. Clin. Invest.* 56:467–75, 1975.
44. Sibbitt WL, Bankhurst AD, Williams RC: Studies in cell subpopulations mediating mitogen hyporesponsiveness in patients with Hodgkin's disease. *J. Clin. Invest.* 61:55–63, 1978.
45. Goodwin JS, Messner RP, Bankhurst AD, Peake GT, Saiki JH, Williams RC: Prostaglandin producing suppressor cells in Hodgkin's disease. *N. Engl. J. Med.* 297:963–8, 1977.
46. Goodwin JS, Bankhurst AD, Messner RP: Suppression of human T cell mitogenesis by prostaglandin: existence of a prostaglandin-producing suppressor cell. *J. Exp. Med.* 146:1719–34, 1977.
47. Bockman RS: Stage-dependent reduction in T colony formation in Hodgkin's disease. *J. Clin. Invest.* 66:523–31, 1980.

48. Pelus C, Bockman R: Increased prostaglandin synthesis by macrophages from tumor-bearing mice. *J. Immunol.* 123:2118–25, 1979.
49. Berlinger NT, Hilal EY, Oettgen HF, Good RA: Deficient cell-mediated immunity in head and neck cancer patients secondary to autologous suppressive immune cells. *Laryngoscope* 88:470–81, 1978. .
50. Jerrells TR, Dean JH, Richardson GL, McCoy JL, Herberman RB: Role of suppressor cells in depression of in vitro lymphoproliferative responses of lung cancer and breast cancer patients. *J. Natl. Cancer Inst.* 61:1001–9, 1978.
51. Herr HW: Suppressor cells in immunodepressed bladder and prostrate cancer patients. *J. Urol.* 123:635–9, 1981.
52. Murray JL, Springle C, Ismael DR, Lee ET, Longley R, Kolmorger G, Nordquist RL: Adherent indomethacin-sensitive supressor cells in malignant melanoma: correlation with clinical studies. *Cancer Immunol. Immunother.* 11:165–72, 1981.
53. Cummings KB, Robertson RP: Prostaglandin: increased production by renal cell carcinoma. *J. Urol.* 118:720–3, 1977.
54. Plescia OJ, Smith AH, Grinwich K: Subversion of the immune system by tumor cells and the role of prostaglandins. *Proc. Natl. Acad. Sci. USA* 72:1848–51, 1975.
55. Grinwich KD, Plescia OJ: Tumor-mediated immunosuppression: prevention by inhibitors of prostaglandin synthesis. *Prostaglandins* 14:1175–82, 1977.
56. Droller MJ, Perlmann P, Schneider MU: Enhancement of natural and antibody-dependent lymphocyte cytotoxicity by drugs, which inhibit prostaglandin production by tumor target cells. *Cell. Immunol.* 39:154–65, 1978.
57. Brunda MJ, Herberman RB, Holden HT: Inhibition of natural killer cell activity by prostaglandins. *J. Immunol.* 124:2682–7, 1980.
58. Prasad KN: Role of prostaglandins in differentiation of neuroblastoma cells in culture. *In*: Powles TJ, Bockman RS, Honn KV, Ramwell P (eds.), *Prostaglandins and Cancer: First International Conference*. Alan R. Liss Inc., New York, pp. 437–51, 1982.
59. Favalli C, Garaci E, Etheredge E, Santoro MG, Jaffe BM: Influence of PGE on the immune response in melanoma bearing mice. *J. Immunol.* 125:89–902, 1980.
60. Stringfellow D, Fitzpatrick F: Prostaglandin D_2 controls pulmonary metastasis of malignant melanoma cells. *Nature* (London) 282:76–80, 1979.
61. Stinnet T, Jaffe BM: Inhibition of B-16 melanoma growth in vitro by prostaglandin D_2. *Prostaglandins* 25:47–54, 1983.
62. Sakai T, Yamaguchi N, Kawai K, Nishino H, Iwashima A: Prostaglandin D_2 inhibits the proliferation of human neuroblastoma cells. *Cancer Lett.* 17:289–94, 1983.
63. Fukushima M, Kato T, Ueda R, Ota K, Narumiya S, Hayaishi O: Prostaglandin D_2: a potential antineoplastic agent. *Biochem. Biophys. Res. Commun.* 105:956–64, 1982.
64. Lynch NR, Salomon J: Tumor growth inhibition and potentiation of im-

munotherapy by indomethacin in mice. *J. Natl. Cancer Inst.* 62:117–21, 1979.
65. Goodwin JS: Prostaglandins and host defense in cancer. *Med. Clins N. Am.* 65:829–44, 1981.
66. Seaman WE, Woodcock J: Human and murine natural killer cell activity may require lipoxygenation of arachidonic acid. *J. Allergy Clin. Immunol.* 74:407–11, 1984.
67. Carine K, Hudig D: Assessment of a role for phospholipase A_2 and arachidonic acid metabolism in human lymphocyte natural cytotoxicity. *Cell Immunol.* 87:270–83, 1984.
68. Bray RA, Brahmi Z: Role of lipoxygenation in human natural killer cell activation. *J. Immunol.* 136:1783–90, 1986.
69. Roder JC, Klein M: Target-effector interaction in the natural killer cell system. *J. Immunol.* 123:2785–90, 1979.
70. Schultz RM, Pavlidis NA, Stylos WA, Chirigos MA: Regulation of macrophage tumoricidal function: a role by prostaglandins of the E series. *Science* 202:320–1, 1978.
71. Murray JL: Prostaglandin modulation of ADCC in human malignant melanoma. *In*: Powles TJ, Bockman RS, Honn KV, Ramwell P (eds.), *Prostaglandins and Cancer: First International Conference.* Alan R. Liss Inc., New York, pp 713–18, 1982.
72. Reddy BS, Cohen LA (eds.): *Diet, Nutrition, and Cancer: A Critical Evaluation.* CRC Press Inc. Boca Raton, 1986.
73. Willett WC, MacMahon B: Diet and cancer: an overview. *N. Engl. J. Med.* 310:697–703, 1984.
74. Horrobin DF: The role of essential fatty acids and prostaglandins in breast cancer. *In*: Reddy BS, Cohen LA (eds.), *Diet, Nutrition, and Cancer, a Critical Evaluation.* CRC Press Inc., Boca Raton, pp. 101–24, 1986.
75. Koivistoinen P, Hyvonen L: Effect of dietary factors on prostanoids. *Ann. Clin. Res.* 16:234–40, 1984.
76. Lands WEM: *Fish and Human Health.* Academic Press, Orlando, 1986.
77. Hopkins GJ, West CE: Possible roles of dietary fats in carcinogenesis. *Life Sci.* 19:1103–6, 1976.
78. Horrobin DF, Manku MS: How do polyunsaturated fatty acids lower plasma cholesterol levels? *Lipids* 18:558–62, 1983.
79. Kagawa Y, Nishizawa M, Suzuki M, Miyatake T, Hamamato T, Gotok, Montonaga E, Izamikawa E, Hirata H, Ebihara A: Eico sapolyenoic acids of serum lipids of Japanese islanders with low incidence of cardiovascular disease. *J. Nutr. Sci. Vitaminol.* 28:441–43, 1982.
80. Panganamala RV, Miller JS, Gwebu ET, Sharma HM, Cornwell DG: Differential inhibitory effects of vitamin E and other antioxidants on prostaglandin synthetase, platelet aggregation and lipoxidase. *Prostaglandins* 14:261–71, 1977.
81. McCay PB, King MM: Vitamin E: its role as a biological free radical scavenger and its relationship to the microsomal mixed fraction oxidase system. *In*: Machlin LJ (ed.), *Vitamin E.* Marcel Dekker Inc., New York, pp. 289–317, 1980.

82. Tappel AL: Vitamin E and selenium protection from in vivo lipid peroxidation. *Ann. N.Y. Acad. Sci.* 355:18–31, 1980.
83. Schmidt K, Moser U: Vitamin C: a modulator of host defense mechanism. *Int. J. Vitamin Nutrit. Res.* 27 (suppl): 363–79, 1985.
84. Cameron E, Pauling L: Ascorbic acid and the glycosaminoglycans. An orthomolecular approach to cancer and other diseases. *Oncology* 27:181–92, 1973.

7

Tissue and organ transplantation

The role of the products of arachidonic acid metabolism in the immunological response to transplanted tissues has been very difficult to evaluate. This is primarily because a large volume of seemingly contradictory data has been generated by experiments in which cyclooxygenase and/or lipoxygenase inhibitors have been used, with quite mixed results, as an adjunct to transplant survival. For example, several investigators have reported that the presence of PGs, particularly PGE, significantly improved skin allograft survival in mice, while others have reported quite the opposite effect (1–4). In a recent paper, however, Foegh et al. have outlined an approach, which helps reconcile such opposing data (5). This is summarized in Figure 7.1. Arachidonic acid metabolites can be considered to belong to one of two groups: those products promoting tissue rejection, and those preventing it. It appears that most, if not all, of these metabolites are present in each transplant, and that survival or rejection of the tissue is dependent upon their balance and net cumulative activity. It is not surprising, therefore, that blanket pharmacological inhibition of cyclooxygenase activity might or might not be successful at prolonging transplant survival. A targeted manipulation of specific metabolites might lead to the generation of more interpretable data.

PGI_2, PGE_2, and PGD_2 are recognized as *antirejection* metabolites. They share an ability to increase cellular cyclic AMP, with a resulting stabilization of inflammatory cells, vascular relaxation, and – in the case of PGE_2 – an ability to suppress immunological recognition and rejection of the transplanted tissue. *Prorejection* metabolites include TXA_2, and lipoxygenase products such as LTB_4 and 5-HETE. These compounds induce both platelet aggregation and vasospasm, activities that, in the case of determining graft survival, eclipse their immunological potential. Much of the activity of this group of compounds appears to be associated with Ca^{2+} influx, and calcium blockers are frequently effective in countering their influence (5).

136

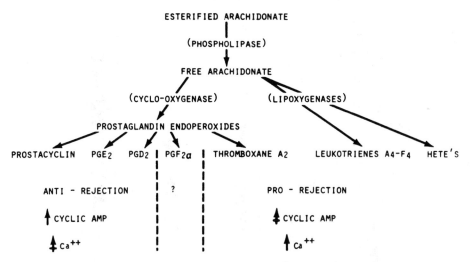

Figure 7.1. The metabolites of arachidonic acid arranged according to their putative roles in transplantation. (Reprinted with permission from Foegh ML, Atijani MR, Helfrich GB, Ramwell PW: Eicosanoids and organ transplantation. *Ann. Clin. Res.* 16:318–32, 1984.)

Generally, PG and LT biosynthesis are accelerated during the allograft response. A dramatic increase in the concentration of blood-borne PGs has been found during allograft rejection (6). This is expected since immunological cell–cell interactions result in the release of diverse arachidonic acid products. During allograft rejection, a variety of cells invade the graft, including monocytes/macrophages, neutrophils, lymphocytes, inflammatory cells, and platelets (7). It is now clear that each of these cells, with the possible exception of the lymphocyte, synthesizes large quantities of arachidonate metabolites. It is thought that lymphocytes, even though they do not seem to synthesize metabolites themselves, can release arachidonic acid in substantial amounts, which is then metabolized by the other cells present in the graft, particularly monocytes/macrophages (8). In addition, Dy et al. found, in a mouse-skin allograft model, that lymphokines released by lymphocytes into the supernatants of mixed lymphocyte cultures (MLC) between donor and recipient could greatly increase the production of PGs (Figure 7.2) by macrophages (9). Since PGs possess a variety of immunological activities (including inhibition of lymphocyte DNA synthesis, lymphocyte cytotoxicity, lymphokine production, and primary antibody response), it is likely that macrophage PG production regulates the immune response within the graft (9).

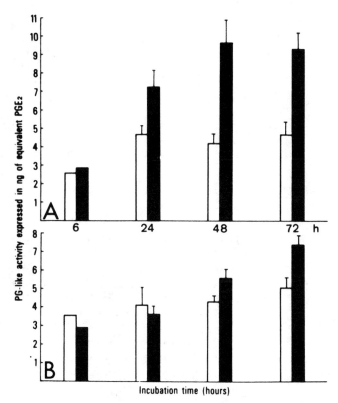

Figure 7.2. (A) Prostaglandin production during mixed lymphocyte culture (MLC) between allograft donor and recipient and (B) in a non-primed MLC. Filled bars indicate PG-like activity in mixed lymphocyte culture supernatants; and open bars indicate PG-like activity in supernatants of single lymphocyte cultures. (Reprinted with permission from Dy M, Debray-Sachs M, Descamps B: Role of macrophages in allograft rejection. *Transpl. Proc.* 9:811–15, 1979.)

In her review of the available literature, Foegh (5) concludes that the treatment of organ allograft recipients with cyclooxygenase inhibitors (with the exception of aspirin (10)) has been disappointing, as these inhibitors prevent the biosynthesis of both antirejection and prorejection prostanoids, and probably shift arachidonic acid metabolism to the generation of lipoxygenase products. Shaw reported that graft survival was not influenced by increasing the supply of PG precursors (Intralipid, Naudicelle), by inhibition of synthesis (BW755C, hydrocortisone), or by the administration of low doses of PGE_1 (11). Graft survival was shortened by the use of one cyclooxygenase inhibitor (Timegadine), and unaffected by the thromboxane synthetase inhibitor, UK-38485, and the thromboxane receptor antagonist, EIP (11). Substitution of the cyclooxygenase inhibitor, ibuprofen, for prednisone in kidney transplant patients resulted in the increased invasion of the graft by inflammatory cells (12). In a study of the survival of skin allografts in mice, Belldegrun et al. showed that low doses of hydrocortisone, indomethacin, and flufenamate were all unable to delay rejection; however, when small doses of hydrocortisone were used in combination with flufenamate or indomethacin, significant graft prolongation was observed (13). Such results have left the issue of PG and LT participation in the immunologic rejection of allograft tissue unresolved to say the least. There is strong evidence from the study of specific metabolites, however, that arachidonic acid metabolism may play an important, if not yet totally defined, role in the allograft response.

ALLOGRAFT PROLONGATION BY PROSTACYCLIN

The accumulation of platelets in transplanted organs undergoing acute rejection has been recognized for many years as a critical part of the rejection process (14). It is also known that prostacyclin (PGI_2) is one of the most powerful inhibitors of platelet aggregation so far discovered (15). The exogenous administration of PGI_2 has been shown to protect organs from hyperacute rejection (16) and the xenograft response (17). Suppression of hyperacute renal allograft rejection in dogs, for example, could be abrogated at least temporarily by PGI_2 infusion; an effect which seems to be related to the inhibition of platelet aggregation (18). Several investigators have reported the successful treatment of hyperacute rejection of human kidney grafts by means of injection of PGI_2 (19,20). Results of these experiments are complicated, however, both by the need for concomitant immunosuppressive therapy, and by the fact that results were sometimes only temporary (20). Shaw recently measured survival of rat cardiac allografts following treatment with only PGI_2 or aspirin during acute rejection, and concluded that both PGI_2 and aspirin appeared to be beneficial to graft survival (10). Interestingly, the benefit of the administration of these two drugs appeared to go beyond platelet inhibition in the grafts, and Shaw hypothesized a direct suppressive effect on cell-mediated immunity through PGI_2 production (10). Suppression of

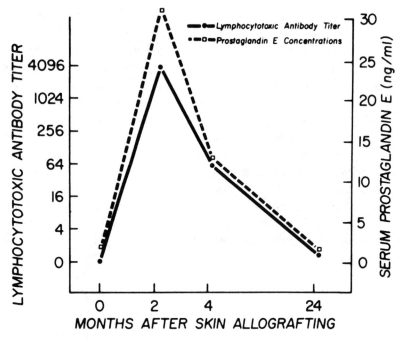

Figure 7.3. Donor-specific serum lymphocytotoxic antibody titers and PGE levels in a dog sensitized by repeated skin allografts from the same specific donor. (Reprinted with permission from Anderson CB, Newton WT, Jaffe BM: Circulating prostaglandin E and allograft rejection. *Transplantation* 19:527–8, 1975.)

cell-mediated immunity by PGI_2 has been confirmed *in vitro* by Leung and Mihich (21), supporting Shaw's interpretation.

Human kidney tissue samples were examined by Leithner et al. in a study of transplant rejection, and it was found that PGI_2 production was significantly enhanced during acute rejection, but not during chronic rejection (22). PGI_2 was actively synthesized, particularly by the renal cortex. Leithner et al. concluded that the increase in PGI_2 generation during acute kidney rejection appeared to be a self-protection mechanism that was, however, overwhelmed during irreversible rejection (22).

PROSTAGLANDIN E_2 AND ALLOGRAFT REJECTION

A relationship between PGE production and allograft rejection was noted as early as 1973, when Loose and DiLuzio reported the prolongation of the survival of murine skin allografts by the administration of exogenous PGE (23). Endogenous PGE activity on the other hand, was shown to play a role in the rejection of rat heart allografts (6) and elevated concentrations were reported approximately two days following the application of canine skin allografts (3). The elevated PGE in this case seemed to correspond with increases in lymphocytotoxic antibody titres as shown in Figure 7.3 (3).

Quagliata et al. studied the effect of E-type PGs on the survival of mouse skin allografts and found that PG administration alone was not enough to prolong graft viability (4). However when PG was administered along with procarbazine, a powerful T-cell depressant, graft survival improved. Quagliata et al. found that the number of B cells, but not T cells, was decreased in the spleens of mice treated with PGE_1 and concluded that the primary inhibitory effect of this metabolite was on B cells (4). Jaffe et al. studied PGE levels following heterotopic rat heart allografting and found that a significant increase in PGE activity occurred between 5 and 7 days post grafting, a time which corresponded with the rejection reaction (24). PGE was most concentrated in the heart grafts themselves, as shown in Figure 7.4. Jaffe et al. hypothesized that the PGE was released as a result of ischemia, and that this release represented a major protective function in the local regulation of blood flow (24,25), or by inhibition of an undefined component of the immune response in the graft. Kakita et al. found, in their study of hamster-to-rat cardiac xenografts, that the injection of PGE_1 prolonged cardiac electrical activity from 73.5 h to 94 h, and suppressed the hemagglutinating antibody titer to hamster red cells at the time of rejection (26).

In 1977, Anderson et al. presented strong evidence for a protective antirejection role for PGE_2 in a murine skin allograft model (2). The effects of 16,16-dimethyl PGE_2 methyl ester (di-M-PGE_2) and indomethacin on the survival mouse skin allografts were studied in B10.D2 female mice receiving skin allografts from (B10.BR × B10.D2) F_1 mice. Daily injections of di-M-PGE_2 increased mean allograft survival from 13.8 days to 16.7 days. Increasing doses of indomethacin were found to correlate inversely with allograft survival; however, this effect could be abrogated by concurrent injection of di-M-PGE_2 (2). These results are summarized in Tables 7.1 and 7.2. Strom and Carpenter tested the ability of 15(S)-15-methyl PGE_1 to act as an antirejection drug in recipients of rat renal allografts and achieved nearly complete protection of the grafts from immunologic damage, even when therapy was withheld until the

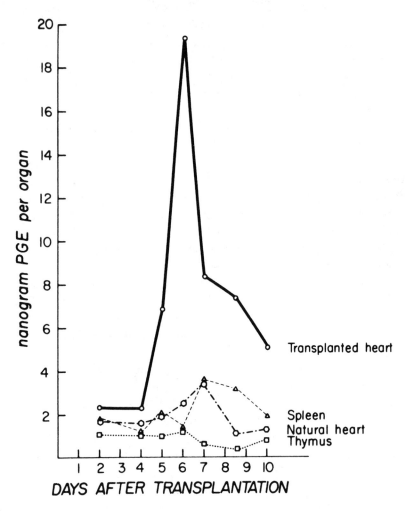

Figure 7.4. Content of PGE in transplanted heart (circles), natural heart (hexagons), spleen (triangles), and thymus (squares) at intervals following rat heart allotransplantation. (Reprinted with permission from Jaffe BM, Moore TC, Vigran TS: Tissue levels of prostaglandin E following heterotopic rat heart allografting. *Surgery* 78:481–4, 1975.)

fourth day posttransplantation (27). 15(S)-15-methyl PGE$_1$ was chosen for this study because of its prolonged biological half-life. Campbell et al. attempted to repeat these observations in a canine renal allograft model, using constant PGE$_1$ administration via implantable pumps, which deliv-

Table 7.1. *Effect of PGE_2 and indomethacin on mouse skin allograft survival.*

Group	No. of animals	Mean skin graft survival (days ± SE)
Control, no injections	23	13.5 ± 0.5
Control, with diluent	36	13.8 ± 0.6
di-M-PGE₂ (5 μg)	32	16.7 ± 0.6
Indomethacin (100 μg) and di-M-PGE₂ (5 μg)	21	16.0 ± 0.6

Note: d-M-PGE₂, dimethyl PGE₂.

Source: Reprinted with permission from Anderson CB, Jaffe BM, Graff RJ: Prolongation of murine skin allografts by prostaglandin E. *Transplantation* 23:444–7, 1977.

Table 7.2. *Effect of indomethacin on the survival of mouse skin allografts and on levels of plasma prostaglandin.*

Group	No.	Mean skin graft survival (days ± SE)	No.	Plasma PGE (pg/ml ± SE)
Control, with diluent	36	13.8 ± 0.6	16	879 ± 80
Indomethacin (100 μg)	39	12.7 ± 0.2	13	717 ± 59
Indomethacin (150 μg)	12	11.8 ± 0.2	–	–
Indomethacin (200 μg)	9	10.9 ± 0.4	9	654 ± 59

Source: Reprinted with permission from Anderson CB, Jaffe BM, Graff RJ: Prolongation of murine skin allografts by prostaglandin E. *Transplantation* 23:444–7, 1977.

ered the PGE_1 into the renal artery of the grafts (28). This attempt failed to demonstrate prolonged allograft survival; however, treated kidneys displayed greatly reduced lymphocyte infiltration, and a greatly increased granulocyte response. In spite of only limited success, Campbell et al. felt that their model represented a potentially important means of manipulating the allograft response at the local level (28).

The release of PGE as a result of the cellular immune response to allografts was studied by Dy et al. by means of mixed lymphocyte cultures (MLC) comprised of donor and recipient lymphocytes (29). While little or no increase in PG production was found in the supernatants of primary MLCs (nonallografted controls), significant PGE_2 was produced by responder cells harvested beginning at 6 days postgrafting and reach-

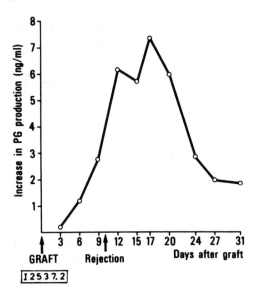

Figure 7.5. Increased PG production during secondary MLC performed at different intervals after grafting. PG was determined using a bioassay. Each point represents the mean value of cells of three mice pooled before culture. (Reprinted with permission from: Dy M, Astoin M, Rigaud M, Hamburger J: Prostaglandin (PG) release in the mixed lymphocyte culture; effect of presensitization by a skin allograft: nature of the PG-producing cell. *Eur. J. Immunol.* 10:121–6, 1980.)

ing a maximum at 17 days postgrafting (Figure 7.5). Dy et al. felt that the increased PGE_2 in the cultures was probably as a result of macrophage production of this metabolite as (a) the cells responsible for the PGE_2 increase could be removed by glass adherence, and (b) production of PGE_2 was inhibited by treatment of spleen cells with silica. They suggested that macrophages were stimulated to produce PGE_2 via lymphokines released during recipient lymphocyte response to donor cells within the graft (29). Such PGE_2 production within the graft would undoubtedly play a role in controlling the magnitude of the antiallograft immune response.

Detecting the onset of such internal immunological control may have prognostic value. Indeed, monitoring the production of PGE_2 within a renal allograft, and the excretion of PGE_2 into the urine have been proposed as early indicators of acute rejection (30). Increases in urine PGE_2

have been found to precede changes in serum creatinine and in creatinine clearance by 1 to 7 days (Figure 7.6).

THROMBOXANE A_2 AS A "PRO-REJECTION" METABOLITE

Like PGE_2, the excretion of TXB_2, a derivative of parent compound TXA_2, into the urine serves as an early indicator of allograft rejection (31). Instead of indicating an ongoing immunoregulatory (protective) response, however, the production of thromboxane signals quite the opposite since TXA_2 is a prorejection metabolite. It seems likely that the balance in the production of prorejection and antirejection metabolites is what helps determine the immunological outcome. Tannenbaum et al. (32) hypothesized that the immunological response to allografted tissue stimulates phospholipase activity, freeing arachidonic acid from storage in tissue lipids. The increase in the intracellular level of free arachidonic acid then leads to a generalized increase in all the products of cyclooxygenase, including thromboxane, PGE_2, and PGI_2. As rejection proceeds, the hypothesis continues, and more immunocompetent cells move into the graft, the generation of lipid hydroperoxides is increased, leading to a selective inhibition of prostacyclin synthetase. Since the other limbs of the prostanoid pathway are not blocked by the presence of these hydroperoxide substances and continue to be produced in large quantities, a significant imbalance between TXA_2 (prorejection) and PGI_2 (antirejection) results in the rejection of the allograft kidney cortex (32). Tannenbaum's hypothesis is summarized in Figure 7.7. Indeed, Tannenbaum et al. report a marked imbalance between PGI_2 and TXA_2 during the early phases of acute rejection (32).

The identification of TXA_2 as a key prorejection metabolite stems primarily from the work of Foegh and her colleagues, and the work of Coffman et al. Foegh et al. found that urinary TXB_2 increased in concentration prior to the rejection of human renal allografts, and could be used as an early indicator of ongoing graft rejection (31). Using a rat heart transplant model, they then determined that the rejection of other organs, besides kidney, also resulted in increased urine TXB_2 (33). Finally, Foegh et al. documented changes in urine TXB_2 in human cardiac transplant patients and found that there was a strong correlation between the occurrence of rejection episodes and high urine TXB_2 (34). They concluded that monitoring the increases in TXB_2 and its meta-

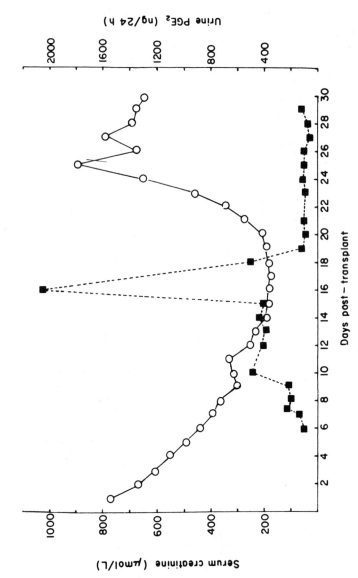

Figure 7.6. Changes in serum creatinine (open circles) and urine PGE$_2$ excretion (filled squares) in a renal transplant patient with an episode of acute rejection. Rejection was diagnosed on day 20. (Reprinted with permission from Thompson D, Rowe DJF, Briggs JD: Measurement of urine prostaglandin E$_2$ as a predictor of acute renal transplant rejection: preliminary findings. *Ann. Clin. Biochem.* 22:161–6, 1985.)

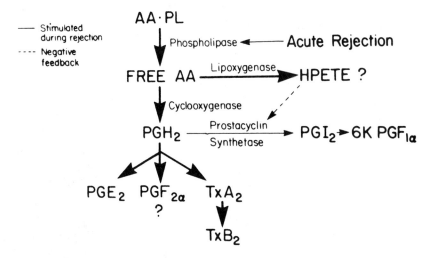

Figure 7.7. Proposed mechanism for the changes in renal arachidonic acid metabolism induced by acute rejection of a renal allograft. AA-PL, phospholipid-bound arachidonic acid; AA, arachidonic acid; 6KPGF$_{1\alpha}$, 6-keto-prostaglandin F$_{1\alpha}$. (Reprinted with permission from Tannenbaum JS, Anderson CB, Sicard GA, McKeel DW, Etheredge EE: Prostaglandin synthesis associated with renal allograft rejection in the dog. *Transplantation* 37:438–43, 1984.)

bolites such as 2,3-dinor-TXB$_2$ served as a reliable immunological monitor of allograft rejection. Increased survival times for rat heterotopic cardiac allografts were observed following the administration of the inhibitors of thromboxane synthetase OXY 1581 (Ono Pharmaceutical, Japan) or L-640,035 (Merck-Frost, Canada) in combination with azathioprine (34).

Coffman et al. transplanted kidneys from Lewis rats to Brown Norway recipients, and monitored histologic rejection and functional decline in graft function (35). As function deteriorated, they found that TXB$_2$ production by *ex vivo*-perfused renal allografts increased progressively, while PGE$_2$ and 6-keto-PGF$_{1\alpha}$ levels remained unchanged. Infusion of the thromboxane inhibitor UK-37248-01 into the renal artery of day three allografts significantly decreased TXB$_2$ excretion and significantly increased renal blood flow and glomerular filtration rate (35).

Because the cellular origin and the immunological significance of TXB$_2$ production has not yet been assessed, the role of this compound in the actual immunological rejection of allografts can only be speculated. It seems likely, however, that the prorejection activities of this metabolite are largely nonimmunological in nature.

LIPOXYGENASEPRODUCTS AND GRAFT REJECTION

Finally, it has been suggested that the production of LTA_4 through LTF_4 as well as HETE production contribute to graft rejection (34). Friedlander et al., however, documented an increase in thromboxane in chronically rejecting kidneys, but not in 12-HETE production (36). These data are significant in that 12-HETE production is considered a reliable marker of platelet activity. The absence of increased 12-HETE suggests a nonplatelet source for TXB_2 in the rejecting kidney (36).

The contribution of the LTs to graft rejection is largely inferred as work in this area is quite limited (34). Recent work, however, implicates at least indirectly the involvement of 5-lipoxygenase products. LTB_4 particularly seems to stimulate Ca^{2+} uptake in PMN, and increase cyclic GMP in lymphocytes, which may play a role in graft rejection.

DIETARY MANIPULATION OF PROSTAGLANDIN PRODUCTION

It is generally recognized that precursor availability is one of the rate limiting steps in eicosanoid production, and that PG production, for example, can be increased by providing increased concentrations of substrate or by shifting precursor entry from one synthetic pathway to another. Santiago-Delpin and Szepsenwol have reported the prolonged survival of skin and tumor allografts in mice supplemented with high-fat diets (37). McHugh et al., reporting a double-blind study of a diet high in polyunsaturated fatty acids on the survival of human cadaveric renal transplants, showed that graft survival during the first 4 months posttransplant was significantly improved by this diet over that of controls (38). By 6 months posttransplant, however, the difference between the groups was no longer significant. Clearly, more studies in this area would be justified.

NATURE'S TRANSPLANT: THE MATERNAL–FETAL
RELATIONSHIP

One of the oldest recognized dilemmas in immunology is how the fetus, which expresses paternal antigens, escapes immunologic rejection by the mother when maternal blood with immunocompetent lymphocytes circulates in contact with the fetal trophoblast. Indeed, it is known that the fetus can resist rejection even in cases where the mother develops anti-HLA antibodies and cytotoxic lymphocytes, or in experimental situations where second set sensitivity to skin grafts of the paternal tissue type can be demonstrated.

One of the earliest theories proposed to explain these observations suggests that the uterus represents an immunologically privileged site, which was exempt from interaction with the immune system. Indeed, it has been shown that allografts placed in the uterus in opposition to a developing trophoblast on the decidual bed survived, whereas grafts at distant sites, or in the nonpregnant horn of a bicornate uterus, were rejected normally (39). It is difficult to explain, however, the existence of extrauterine pregnancy in humans if this is the primary reason the fetus survives.

Likewise the theory that the trophoblast does not express MHC products does not explain all available data. For example, many groups have demonstrated the expression of both parental H-2 haplotypes on murine trophoblasts (40,41). It has been suggested, however, that these antigens may at least be partially blocked by maternal antipaternal antibodies (42). Faulk and colleagues have suggested that the HLA antigens of human trophoblasts are masked by either transferrin or other placental substances (43,44). A similar masking of transplantation antigens expressed by the placenta is currently under investigation. Again, however, it is known that embolization and destruction of trophoblasts in the lungs leads to HLA sensitization of the maternal immune system (45), making this possible explanation also seem incomplete.

The most likely explanation for the survival of the fetus involves the alteration of maternal immune responsiveness. Such an explanation must account for maintenance of the normal maternal immune competence against malignancy, infection, and experimental grafts during pregnancy. Two mechanisms have been suggested, which fit these criteria, and these mechanisms may not be mutually exclusive (45):

1. the local release of immunosuppressive substances in the placenta having nonspecific effects on maternal cellular immunity (ideally, substances hav-

ing a short half-life in the circulation or subject to consumption at the maternal–placental interface); or

2. the presence of specific antipaternal immunosuppressors (such as blocking antibody against paternal haplotypes, antiidiotypic immunity directed against cells or cytotoxic antipaternal effectors, or specific T lymphocyte or macrophage suppressor cells).

It is now clear that hormones such as progesterone serve a critical role in down regulating the local maternal immune response to paternal antigens during pregnancy (45,46). Furthermore, progesterone exerts at least part of its activity by means of regulating PG synthesis (45,47), and there is increasing evidence that the products of arachidonic acid may be instrumental in determining the immunological survival of the trophoblast and the fetus. T-lymphocyte suppressor cells, which can inhibit both T- and B-cell proliferation, have been identified in human newborns (48,49). Monocyte suppressors of T-cell activity have also been demonstrated, and it has been reported that approximately 60% of their activity is mediated through production of PGE_2 (50). Preliminary results of Fischer et al. indicate that this PGE_2 may exert its inhibitory activity on lymphocyte proliferation via activation of a short-lived T-suppressor cell (51), which may in turn act as a feedback inhibitor of suppressor monocytes, a cycle of potential importance to fetal survival (50).

The human fetus exists in an environment rich in PGs. For example, a study of human fetal plasma, obtained by fetoscopy at 16–20 weeks gestation, demonstrated the presence of significant quantities of PGE_2, $PGF_{2\alpha}$, and 6-keto-$PGF_{1\alpha}$, a nonenzymatically formed product of PGI_2 degradation (52). It is of interest that PGI_2 concentrations in fetal plasma were greater than observed in maternal peripheral plasma. This was particularly true of the PGI_2 product, 6-keto-$PGF_{1\alpha}$, suggesting an important role for this antirejection metabolite during early pregnancy. Concentrations of PGE_2, $PGF_{2\alpha}$, and 13,14-dihydro-15-keto-$PGF_{2\alpha}$ in umbilical plasma were also greater than those in maternal plasma (Table 7.3) (53). The concentrations of these three metabolites were all significantly increased in umbilical plasma after the onset of labor, suggesting that labor is a stimulus to PG production by the fetoplacental unit (54). Unfortunately, little information is currently available concerning the lipoxygenase products of arachidonic acid metabolism in fetal and neonatal tissues.

Likewise, uterine and intrauterine tissues also produce PGs. The amnion has been shown to be a significant source of PGE_2 (another antire-

Table 7.3. *PG concentrations in maternal and umbilical circulations at full-term pregnancy.*

	Maternal circulating		Umbilical (spontaneous labor)	
	Late pregnancy n = 13	Labor; cervix, 5—8 cm n = 5	Artery n = 12	Vein n = 12
PGE	4.8 ± 1.0	5.4 ± 2.2	109.3 ± 26.9	241.9 ± 24.9
PGF	6.2 ± 0.5	12.4 ± 3.5	79.7 ± 10.4	87.8 ± 11.1
PGFM	59.0 ± 7.8	282.7 ± 55.3	639.9 ± 180.2	630.8 ± 107.3

Note: All values in the table are expressed as mean pg/mL ± SEM.

Source: Reprinted with permission from Mitchell MD, Brennecke SP, Saeed SA, Strickland DM: Arachidonic acid metabolism in the fetus and neonate. *In*: Cohen MM (ed.), *Biological Protection with Prostaglandins*, Vol 1. CRC Press, Boca Raton, pp 27–44, 1985.

jection metabolite), and also TXB_2 (a prorejection metabolite) (55,56). The rate of PG synthetase activity in the uterus appears to increase at the onset of labor (57), as does the concentration of PGE_2 (56). The cervix also appears to be a major source of PG, particularly of the E series, and it has been hypothesized that the softening and dilation of the cervix are dependent upon locally formed PGs (58). Uterine and intrauterine tissues also produce lipoxygenase products of arachidonic acid metabolism. Human amnion, decidua vera, and placenta synthesize 12-HETE, and a small amount of 5-HETE (59). The chorion, however, produces only a trace amount of 12-HETE. It has been hypothesized that since the HETEs are powerful chemotactic agents for human neutrophils, eosinophils, and macrophages, the production of these metabolites may regulate leukocyte and/or macrophage infiltration during pregnancy (60).

FETAL PROSTAGLANDIN-E_2-PRODUCING CELLS

It has been shown that the newborn circulation contains PGE_2-producing suppressor cells, which are either monocytes or OKT4(+) lymphocytes (61,62). These cells may be instrumental in assuring the immunological survival of the fetus, since maternal cells appear to be quite sensitive to PGE_2-mediated suppression while fetal cells are not (63–65). In fact, it requires approximately 100 times the concentration of PGE_2, which inhibits maternal cells (1.4×10^{-8} M) to suppress the response of newborn cells $(1.4 \times 10^{-6}$ M) (64). Papadogiannakis et al. found that the target of PGE_2-mediated suppression in this system was the highly sensitive maternal OKT4(+) T-cell subset, and that fetal OKT4(+)8(−) T cells were resistant (63). Taken together, these data suggest that the fetus may create an environment, rich in PGE_2, in which its own OKT4(+) cells are functional, but maternal cells of the same phenotype are not. In this way, the fetus may be protected against maternal T cells, which occasionally cross the placenta during pregnancy (66,67).

From a review of arachidonic acid metabolism as it relates to the maternal-fetal relationship (68), it is clear that the balance between the various metabolites (concentration, rates of production, turnover, etc.) is critical in determining their net effect on the entire fetal development and the birth process. The immunological relationship between mother and fetus, likewise, undoubtedly depends on this balance and deserves the research attention it is now receiving.

LITERATURE CITED

1. Pelus LM, Strausser HR: Prostaglandins and the immune response. *Life Sci.* 20:903–4, 1977.
2. Anderson CB, Jaffe BM, Graff RJ: Prolongation of murine skin allografts by prostaglandin E. *Transplantation* 23:444–7, 1977.
3. Anderson CB, Newton WT, Jaffe BM: Circulating prostaglandin E and allograft rejection. *Transplantation* 19:527–8, 1975.
4. Quagliata F, Lawrence VJW, Phillips-Quagliata JM: Prostaglandin E as a regulator of lymphocyte function. *Cell. Immunol.* 6:457–65, 1973.
5. Foegh ML, Alijani MR, Helfrich GB, Ramwell PW: Eicosanoids and organ transplantation. *Ann. Clin. Res.* 16:318–32, 1984.
6. Moore TC, Jaffe BM: Prostaglandin E levels of heterotopic rat heat allografts and host lymphoid tissues at intervals post-grafting. *Transplantation* 18:383–5, 1974.
7. Hayry P: Intragraft events in allograft destruction. *Transplantation* 38:1–6, 1984.
8. Goldyne ME, Stobo JD: T-lymphocytes, a source of arachidonic acid for the synthesis of eicosanoids by human monocytes/macrophages. *Adv. Prostgl. Thrombox. Leukotr. Res.* 12:39–43, 1983.
9. Dy M, Debray-Sachs M, Descamps B: Role of macrophages in allograft rejection. *Transpl. Proc.* 9:811–15, 1979.
10. Shaw JFL: Prolongation of rat cardiac allograft survival by treatment with prostacyclin or aspirin during acute rejection. *Transplantation* 35:526–9, 1983.
11. Shaw JFL, Greatorex RA: Drugs affecting the prostaglandin synthetic pathway and rat heart allograft surival. *Adv. Prostgl. Thrombox. Leukotr. Res.* 13:219–20, 1985.

12. Kreis H, Chekoff N, Droz D, Noel LH, Tolani M, Descamps JM, Chateroid L, Lacombe M, Crosnier J: Nonsteroid antiinflammatory agents as a substitute treatment for steroids in ATG-AM treated cadaver kidney recipients. *Transplantation* 37:139–45, 1984.

13. Belldegrun A, Cohen IR, Frenkel A, Servadio C, Zor U: Hydrocortisone and inhibitors of prostaglandin synthesis. *Transplantation* 31:407–8, 1981.

14. Burrows L, Haimov M, Aldedort L: The platelet in the obliterative vascular rejection phenomenon. *Transpl. Proc.* 5:157–60, 1973.

15. Moncada S, Gryglewski R, Bunting S, Vane JR: An enzyme isolated from arteries transforms prostaglandin endoperoxides to an unstable substance that inhibits platelet aggregation. *Nature* 263:663–5, 1976.

16. Mundy AR, Bewick M, Moncada S, Vane JR: Short term suppression of hyperacute renal allograft rejection in presensitized dogs with prostacyclin. *Prostaglandins* 19:595–603, 1980.

17. Mundy AR: Prolongation of cat to dog renal xenograft survival with prostacyclin. *Transplantation* 30:226–8, 1980.

18. Moncada S, Vane JR: Short term suppression of hyperacute renal allograft rejection in presensitized dogs with prostacyclin. *Prostaglandins* 19:595–603, 1980.

19. Sinzinger H, Leithner C, Schwarz M: Monitoring of human kidney transplants using quantification of autologous 111 indium-oxine platelet label deposition: beneficial effect of PGI_2 treatment in acute and chronic rejection. *Thromb. Haemost.* 46:263–8, 1981.

20. Leithner C, Sinzinger H, Pohanka E, Schwarz M, Syre G: 111-Indium-markierte Thrombozyten in der Diagnostik von akuten Nierentransplantatabstossugen zur Observierung der Therapie von Abstossungsreaktionen mit Prostacyclin. *Klin. Wochenschr.* 96:112–17, 1984.

21. Leung KH, Mihick E: Prostaglandin modulation of development of cell-mediated immunity in culture. *Nature* (London) 288:597–600, 1980.

22. Leithner C, Sinzinger H, Silberbauer K, Wolf A, Stummvoll HK, Pinggera W: Enhanced prostacyclin synthesis in acute human kidney transplant rejection. *Proc. Eur. Dial. Transpl. Assoc.* 17:424–8, 1980.

23. Loose LD, DiLuzio NR: Effect of prostaglandin E_1 on cellular and humoral immune responses. *J. Reticuloendothel. Soc.* 13:70–7, 1973.

24. Jaffe BM, Moore TC, Vigran TS: Tissue levels of prostaglandin E following heterotopic rat heart allografting. *Surgery* 78:481–4, 1975.

25. McGiff JC, Growshaw K, Terragno NA, Lanigro AJ, Strand JC, Williamson MA, Lee JB, Ng KKF: Prostaglandin-like substances appearing in canine renal venous blood during renal ischemia. Their partial characterization by pharmacologic and chromatographic procedures. *Circ. Res.* 27:765–82, 1970.

26. Kakita A, Blanchard J, Fortner JG: Effectiveness of prostaglandin E_1 and procarbazine hydrochloride in prolonging survival of vascularized cardiac hamster-to-rat xenograft. *Transplantation* 20:439–42, 1975.

27. Strom TB, Carpenter CB: Prostaglandin as an effective antirejection therapy in rat renal allograft recipients. *Transplantation* 35:279–81, 1983.

28. Campbell D, Wiggins R, Kunkel S, Juni J, Tuscan M, Shapiro B, Niederhuber

J: Constant intrarenal infusion of PGE$_1$ into a canine renal transplant using a totally implantable pump. *Transplantation* 38:209–12, 1984.

29. Dy M, Astoin M, Rigaud M, Hamburger J: Prostaglandin (PG) release in the mixed lymphocyte culture; effect of presensitization by a skin allograft: nature of the PG-producing cell. *Eur. J. Immunol.* 10:121–6, 1980.

30. Thompson D, Rowe DJF, Briggs JD: Measurement of urine prostaglandin E$_2$ as a predictor of acute renal transplant rejection: preliminary findings. *Ann. Clin. Biochem.* 22:161–5, 1985.

31. Foegh ML, Alijani M, Helfrich GB, Schreiner GE, Ramwell PW: Urine thromboxane as an immunological monitor in kidney transplant patients. *Transplant. Proc.* 16:1603–5, 1984.

32. Tannenbaum JS, Anderson CB, Sicard GA, McKeel DW, Etheredge EE: Prostaglandin synthesis associated with renal allograft rejection in the dog. *Transplantation* 37:438–43, 1984.

33. Foegh ML, Alijani M, Khirabadi BS, Shapiro R, Goldman MH, Lower RR, Ramwell PW: Monitoring rat heart allograft rejection by urinary thromboxane. *Transpl. Proc.* 16:1606–8, 1984.

34. Foegh ML, Alijani MR, Helfrich GB, Khirabadi BS, Goldman MH, Lower RR, Ramwell PW: Thromboxane and leukotrienes in clinical and experimental transplant rejection. *Adv. Prostagl. Thrombox. Leukotr. Res.* 13:209–17, 1985.

35. Coffman TM, Yarger WE, Klotman PE: Functional role of thromboxane production by acutely rejecting renal allografts in rats. *J. Clin. Invest.* 75:1242–8, 1985.

36. Friedlander G, Moulonguet-Doleris L, Kourilsky O, Nussaume O, Ardaillou R, Sraer JD: Prostaglandin synthesis by glomeruli isolated form normal and chronically rejected human kidneys. *Contr. Nephrol.* 41:20–2, 1984.

37. Santiago-Delpin EA, Szepsenwol J: Prolonged survival of skin and tumor allografts in mice on high-fat diets. *J. Natl. Cancer Inst.* 59:459–61, 1977.

38. McHugh MI, Wilkinson R, Elliot RW, Field EJ, Dewar P, Hall RR, Taylor RMR, Uldall AR: Immunosuppression with polyunsaturated fatty acids in renal transplantation. *Transplantation* 24:263–7, 1977.

39. Beer AE, Billingham RE: Host responses to intrauterine tissue, cellular and fetal allografts. *J. Reprod. Fertil. Suppl.* 21:59–88, 1974.

40. Chatterjee-Hasrouni S, Lala PK: The localization of H-2 antigens in mouse trophoblast cells. *J. Exp. Med.* 149:1238–53, 1979.

41. Pavia CG, Stites DP, Fraser R: Transplantation antigenic expression on murine trophoblast. *Cell. Immunol.* 64:162–76, 1981.

42. Voisin GA, Chaouat G: Demonstration, nature, and properties of maternal antibody fixed on placenta and directed against paternal antigens. *J. Reprod. Fertil.* 21 (suppl.):89–103, 1974.

43. Faulk WP, Galbraith GMP: Trophoblast transferrin and transferrin receptors in the host-parasite relationship of human pregnancy. *Proc. R. Soc. Lond. B Biol. Sci.* 204:85–97, 1979.

44. McIntyre JA, Faulk WP: Trophoblast modulation of maternal allogeneic recognition. *Proc. Natl. Acad. Sci. USA* 76:4029–32, 1979.

45. Siiteri PK, Stites DP: Immunologic and endocrine interrelationships in pregnancy. *Biol. Reprod.* 26:1–14, 1982.
46. Moriyana I, Sugawa T: Progesterone facilitates implantation of xenogeneic cultured cells in hamster uterus. *Nat. New Biol.* 236:150–2, 1972.
47. Lewis DA, Symons AM, Ancill RJ: The stabilization-lysis action of anti-inflammatory steroids on lysosomes. *J. Pharm. Pharmacol.* 22:902–8, 1970.
48. Hayward AR, Lydyard PM: Suppression of B lymphocyte differentiation by newborn T lymphocytes with an Fc receptor for IgM. *Clin. Exp. Immunol.* 34:374–8, 1978.
49. Durandy A, Fischer A, Griscelli C: Active suppression of B lymphocyte maturation by two different newborn T lymphocyte subsets. *J. Immunol.* 123:2644–50, 1979.
50. Fischer A, Durandy A, Mannas S, McCall E, Dray F, Griscelli C: Lack of prostaglandin E_2-mediated monocyte suppressive activity in newborn and mothers. *Clin. Exp. Immunol.* 49:377–85, 1982.
51. Fischer A, Durandy A, Griscelli C: Role of PGE_2 in the activation of non-specific T suppressor lymphocytes. *J. Immunol.* 126:1452–8, 1981.
52. Mackenzie IZ, MacLean DA, Mitchell MD: Prostaglandins in the human fetal circulation in mid-trimester and term pregnancy. *Prostaglandins* 20:649–54, 1980.
53. Mitchell MD, Flint APF, Bibby J, Brunt J, Arnold JM, Anderson ABM, Turnbull AC: Plasma concentrations of prostaglandins during late human pregnancy: influence of normal and pre-term labor. *J. Clin. Endocrinol. Metab.* 46:947–54, 1978.
54. Mitchell MD, Brunt J, Bibby J, Flint APF, Anderson ABM, Turnbull AC: Prostaglandins in the human umbilical circulation at birth. *Br. J. Obstet. Gynaecol.* 85:114–8, 1978.
55. Mitchell MD, Bibby JG, Hicks BR, Redman CWG, Anderson ABM, Turnbull AC: Thromboxane B_2 and human parturition: concentrations in the plasma and production in vitro. *J. Endocrinol.* 78:435–41, 1978.
56. Mitchell MD, Bibby J, Hicks BR, Turnbull AC: Specific production of prostaglandin E_2 by human amnion in vitro. *Prostaglandins* 15:377–82, 1978.
57. Okazaki T, Casey ML, Okita JR, MacDonald PC, Johnston JM: Initiation of human parturition. XII. Biosynthesis and metabolism of prostaglandins in human fetal membranes and uterine decidua. *Am. J. Obstet. Gynecol.* 139:373–81, 1981.
58. Ellwood DA, Mitchell MD, Anderson ABM, Turnbull AC: In vitro production of prostanoids by the human cervix during pregnancy: preliminary observations. *Br. J. Obstet. Gynaecol.* 87:210–14, 1980.
59. Saeed SA, Mitchell MD: Formation of arachidonate lipoxygenase metabolites by human fetal membranes, uterine decidua vera, and placenta. *Prostagl. Leukotr. Med.* 8:635–40, 1982.
60. Mitchell MD, Strickland DM, Brennecke SP, Saeed SA: New aspects of arachidonic acid metabolism and human parturition. *In:* Porter JC (ed.), *Proceedings of the Ross Conference on Obstetric Research, Initiation of Parturition: Prevention of Prematurity.* Ross Inc., Columbus, 1983.

61. Johnsen SA, Olding LB, Westberg N, Wilhelmsson L: Strong suppression by mononuclear leukocytes from human newborns as maternal leukocytes: mediation by prostaglandins. *Clin. Immunol. Immunopathol.* 23:606–15, 1982.
62. Pirquet PF, Irle C, Vassalli P: Immunosuppressor cells from newborn mouse spleen are macrophages differentiating in vitro from monoblastic precursors. *Eur. J. Immunol.* 11:56–61, 1981.
63. Papadogiannakis N, Johnsen SA, Olding LB: Human fetal/neonatal suppressor activity: relation between OKT phenotypes and sensitivity to prostaglandin E_2 in maternal and neonatal lymphocytes. *Am. J. Reprod. Immunol. Microbiol.* 9:105–10, 1985.
64. Papadogiannakis N, Johnsen SA, Olding LB: Strong prostaglandin associated suppression of the proliferation of human maternal lymphocytes by neonatal lymphocytes linked to T versus T cell interactions and differential PGE_2 sensitivity. *Clin. Exp. Immunol.* 61:125–34, 1984.
65. Johnsen SA, Olofsson A, Green K, Olding LB: Strong suppression by mononuclear leukocytes from cord blood of human newborns on maternal leukocytes associated with differences in sensitivity to prostaglandin E_2. *Am. J. Reprod. Immunol.* 4:45–9, 1983.
66. Clark DA: Prostaglandins and immunoregulation during pregnancy. *Am. J. Reprod. Immunol. Microbiol.* 9:111–12, 1985.
67. Hunziker RD, Gambel P, Wegman TG: Placenta as a selective barrier to cellular traffic. *J. Immunol.* 133:667–71, 1984.
68. Mitchell MD, Brennecke SP, Saeed SA, Strickland DM: Arachidonic acid metabolism in the fetus and neonate. *In*: Cohen MM (ed.), *Biological Protection with Prostaglandins*, Vol 1. CRC Press, Boca Raton, pp. 27–44, 1985.

8

Rheumatoid arthritis and autoimmunity

Rheumatoid arthritis is a chronic inflammatory disease involving the joints and connective tissues throughout the body. The disease is most common in females between the ages of 30 and 40 years, and has a generally unpredictable course. The etiology of rheumatoid arthritis is still unknown; however, there is evidence that immune complexes are involved in the pathogenesis of the synovial lesions characteristic of the disease. These antigen–antibody complexes can activate complement, a process which releases chemotactic factors and attracts granulocytes. The granulocytes release lysosomal enzymes and vasoactive amines and cause inflammatory tissue destruction. Complexes consisting of cell nuclear fragments and anti-DNA antibodies also contribute to joint inflammation. Antinuclear factors are often present in the joint fluid but absent in the serum. It appears that other diseases, such as gout, spondylitis, psoriatic arthritis, colitic arthritis, and Reiter's syndrome share some of these characteristics (1).

Another hallmark of rheumatoid arthritis is the development and circulation of rheumatoid factor, which is antibody to IgG. Rheumatoid factor includes both 7S and 19S immunoglobulins with specificity to certain determinants on the IgG molecule, determinants thought to be exposed by physical aggregation of IgG or immune-complex formation (2). Rheumatoid factor can be detected in the serum of approximately 80% of patients with rheumatoid arthritis (3) and can also be seen in many nonrheumatoid diseases including viral and parasitic infections, tuberculosis, leprosy, and subacute bacterial endocarditis (4). In addition, the majority of renal transplant recipients develop rheumatoid factor, as do patients receiving multiple transfusions (5,6). The development of rheumatoid factor, therefore, seems to be a manifestation of chronic antigen stimulation (7). In arthritis, rheumatoid factor interacts with antigen–antibody complexes as well as with aggregates of IgG, fixes complement, and initiates a destructive Arthus-reaction-like inflamma-

157

tion in the joint space. Again, the attraction of granulocytes and their release of lysosomal enzymes results in proteolytic destruction of synovial tissues (8).

Harris et al. demonstrated collagenase activity in the synovial fluid of some patients (9,10). These investigators published histological evidence that a diffuse, proliferative synovitis results from collagenase activity. Synovitis is a dominant feature of early lesions of rheumatoid arthritis. As the disease progresses, the collagenase activity and proliferation continues, and structures such as tendon and cartilage are broken down.

In addition to the presence of proteolytic enzymes, Robinson et al. demonstrated that PGs are produced in large amounts by rheumatoid synovial fragments and can be readily detected in the synovial fluid (11). The generation of PGs as part of the rheumatoid lesion is important since these metabolites appear to be active in the synovial inflammation as well as in bone resorption through the stimulation of osteoclasts (12), and immunological depression in patients with advanced disease (13).

PROSTAGLANDINS AND RHEUMATOID SYNOVIUM

Krane and Dayer et al. studied the behavior of explanted cultures of synovial cells in an effort to determine the origin of both collagenase and PG in the synovial fluid (12,14). They found that variable percentages of dispersed synovial cells adhered to the culture vessel surface, and that these adherent cells were responsible for the production of all of the measurable collagenase and most of the PG in the cultures (14). Most of the adherent cells were quite large (20–30 μm in diameter) and had a characteristic stellate form. These cells did not bear Fc receptors or any other characteristic macrophage markers, did not produce lysozyme, and did not have an ultrastructural architecture characteristic of macrophages (15). The conclusive identity of this predominant cell type is yet to be determined; however, similiar results have been reported by Krakaner and Zurier (16) and Woolley et al. (17), who have identified the cells as dendritic cells.

Dayer et al. found that when monocytes were cocultured with the adherent synovial cell population, there was a significant increase in collagenase and PGE_2 production (18,19). This is illustrated in Figure 8.1. The stimulation was roughly proportional to the number of monocytes added to the cultures. In addition, it was found that the same stimulation of collagenase and PG synthesis could be accomplished by the addition of cell-free conditioned medium from mononuclear cell cul-

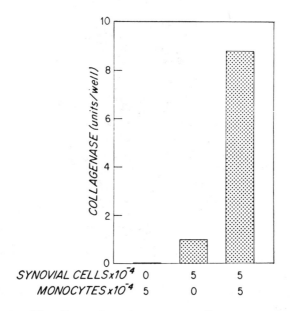

| SYNOVIAL CELLS x10⁻⁴ | 0 | 5 | 5 |
| MONOCYTES x10⁻⁴ | 5 | 0 | 5 |

Figure 8.1. The effects of cocultivation on collagenase production by adherent rheumatoid synovial cells and peripheral blood monocytes. Cells were incubated in serum-containing media 3 days prior to assays for latent collagenase. (Reprinted with permission from: Krane SM: Aspects of the cell biology of the rheumatoid synovial lesion. *Ann. Rheumatic Dis.* 40:433–48, 1981.)

tures, to the synovial cell cultures (18,19). Experiments using purified mononuclear cell populations showed that the monocyte/macrophage was responsible for these effects and that lymphocytes played no role in stimulating collagenase and PGE_2 production (15). The monocyte/macrophage product with stimulating activity has been characterized and has been labeled mononuclear cell factor, or MCF (20). MCF production by the monocyte was greatly enhanced by the presence of aggregated IgG or Fc fragments (15).

Cell–cell contact does not appear to be necessary for the stimulation of PGE_2 production by synovial cells, since incubation of these cells with monocyte-free MCF results in increased PGE_2 synthesis. The addition of indomethacin during incubation has shown that MCF also acts directly to increase the sensitivity of synovial cells to PGE_2, presumably by increasing the number of PGE_2 receptors or altering binding affinity (12). These observations are particularly interesting since MCF appears to be chemically similar, if not identical to, interleukin-1 (IL-1) (12,21,22).

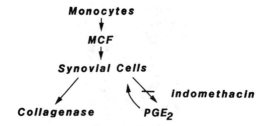

Figure 8.2. Interrelationship of MCF and PGE$_2$ production by synovial cells. The increased levels of PGE$_2$ produced by synovial cells in response to MCF act back on the synovial cells. Indomethacin, by blocking cyclooxygenase and decreasing PGE$_2$ synthesis, can unmask certain effects of MCF on synovial cells, such as increasing cyclic AMP response to PGE$_2$ and stimulating cell division. (Reprinted with permission from Krane SM: Aspects of the cell biology of the rheumatoid synovial lesion. *Ann. Rheumatic Dis.* 40:433–88, 1981.)

MCF derived from monocytes appears to control DNA synthesis and, therefore, proliferation of several types of cells (12). For example, the addition of increasing amounts of MCF results in decreased uptake of [³H]thymidine by synovial cells, an observation thought to be as a result of MCF-induced stimulation of endogenous PGE$_2$ synthesis, since indomethacin reverses MCF suppression of proliferation (20). This relationship is shown in Figure 8.2. Likewise mononuclear-cell-conditioned medium suppresses fibroblast proliferation through the stimulation of PGE$_2$ synthesis (23). Again, growth inhibition is reversed by the addition of cyclooxygenase inhibitors (23).

PGs synthesized by synovial cells and fibroblasts in culture can, in turn, affect additional cellular functions. Collagenase production appears to be controlled through changes in PGE$_2$ concentration (12), although conflicting data show that inhibition of PG synthesis does not alter the release of this enzyme (24). Krane has explained this difference in results as evidence of increased sensitivity of rheumatoid synovial cell cultures to PGE$_2$ with the progression of the disease (12).

PROSTAGLANDIN PARTICIPATION IN IMMUNE CHANGES

In their 1981 paper, Cueppens et al. reported that IgM rheumatoid factor production by cultured lymphocytes, isolated from the peripheral blood of patients with rheumatoid arthritis, could be inhibited by the addition of any of several nonsteroidal antiinflammatory agents including indomethacin, carprofen, and piroxicam (25). These authors suggest that such inhibition of rheumatoid factor production by NSAIA treatment during the early rheumatoid inflammatory process may represent a previously unrecognized mode of action of the drugs. Such results also confirm the direct role of PGs in the generation of rheumatoid factor. In a later study, Goodwin et al. found that prolonged patient treatment with piroxicam resulted in a long-term and significant reduction in IgM rheumatoid factor levels (to 89% of baseline at 4 weeks, 79% of baseline at 6 weeks, 68% of baseline at 8 weeks, and 62% of baseline at 10 weeks following the initiation of piroxicam treatment) with no drop in patient serum IgM concentration (26). These results are shown in Figure 8.3.

It had not previously been explained why rheumatoid factors are preferentially produced in the involved joints of patients with rheumatoid arthritis (27). Goodwin's results suggested that the observation that PGE_2 levels in rheumatoid synovial fluid can be three orders of magnitude greater than concentrations found in the circulation or in normal tissues (28) is directly related to rheumatoid factor production in these locations.

Altered interleukin-2 (IL-2) production and response has also been reported in patients with rheumatoid arthritis (29), an effect which may also be PG based. Using an IL-2-dependent cell line, McKenna et al. showed that lymphocytes from patients with active rheumatoid arthritis displayed reduced IL-2 production when compared to patients with inactive disease (30). It is interesting to note that the patients with inactive disease were receiving both NSAIA and gold therapy. As discussed in Chapter 4, it appears that IL-2 production is indirectly regulated, through PGE_2 activation of a CD8(+) lymphocyte subset. A depression in IL-2 production in patients with rheumatoid arthritis would likely manifest itself as a reduction in cell-mediated immunity.

While studies by Waxman et al. (31) and Lockshin et al. (32) raised some early doubts, the general depression of the cell-mediated immune response as a result of rheumatoid arthritis and other chronic inflammatory disease is now recognized. Silverman et al., for example, have shown that peripheral blood mononuclear cells from patients with rheumatoid arthritis displayed a depressed response *in vitro* to PHA over a range of mitogen concentrations (33). Wolinsky et al. also observed this depression in the PHA response of arthritic patients (13) and found that the depressed response could be partially reversed by inhibition of PG syn-

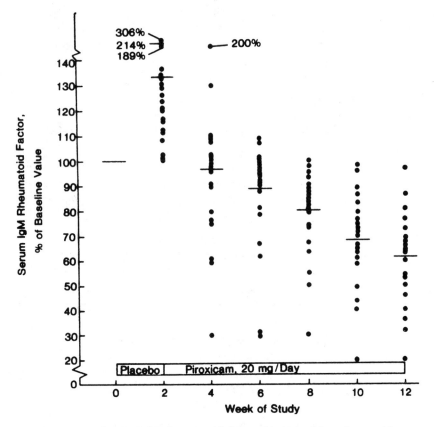

Figure 8.3. Serum IgM rheumatoid factor levels in 20 patients with rheumatoid arthritis while receiving various nonsteroidal antiinflammatory agents, after 2 weeks of placebo medication, and after 2,4,6,8, and 10 weeks of piroxicam administration. All values are expressed as a percentage of baseline level at week 0. Mean rheumatoid factor levels increased 33% after 2 weeks of placebo administration. These levels dropped with piroxicam administration; after 4,6,8, and 10 weeks of piroxicam therapy, rheumatoid levels were below baseline. (Reprinted with permission from Goodwin JS, Ceuppens JL, Rodrigues MA: Administration of nonsteroidal antiinflammatory agents in patients with rheumatoid arthritis. *J. Am. Med. Ass.* 250:2485–88, 1983.)

thetase with NSAIA (26). The response of lymphocytes isolated from rheumatoid arthritis patients treated with piroxicam is summarized in Figure 8.4.

Lymphocytes from rheumatoid patients appear to be more sensitive to the inhibitory effects of exogenous PGE_2 than normal lymphocytes (13).

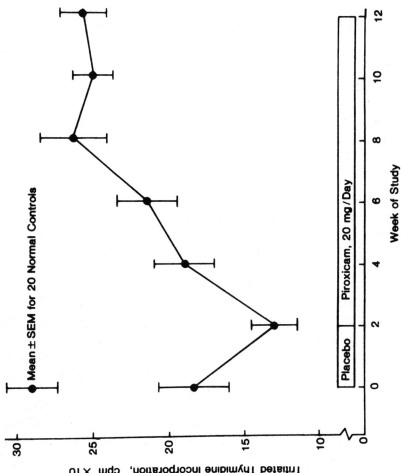

Figure 8.4. Response of lymphocytes from patients with rheumatoid arthritis to PHA (2 mg/L) while receiving various nonsteroidal antiinflammatory agents, after 2 weeks of placebo medication, and after piroxicam administration. Results represent mean ± SEM from data on 20 patients. Baseline (week 0) response was depressed compared to that of 20 normal controls. It fell further after placebo therapy and then gradually increased during piroxicam administration. Similar results were obtained with two other concentrations of PHA. (Reprinted with permission from Goodwin JS, Ceuppens JL, Rodriguez MA: Administration of nonsteroidal antiinflammatory agents in patients with rheumatoid arthritis. *J. Am. Med. Ass.* 250:2485–8, 1983.)

Wolinsky et al. felt that this might explain why NSAIA did not more completely reverse their depressed PHA response (13). In such a scenario, it would take far less PGE_2 to produce suppression in a sensitive cell population than normal.

Patients with active rheumatoid arthritis have fewer than normal, phenotypically identifiable suppressor cells in their peripheral blood (34,35) and an increase in the proportion of null cells (35). Goodwin et al. studied the effect of administration *in vivo* of NSAIA on peripheral blood concentrations of T-helper cells, T-suppressor cells, and null cells in patients with active disease (26). They found that OKT8-staining cells rose from subnormal levels to significantly higher than normal as a result of 12 weeks of piroxicam treatment. Levels of OKM1(+) null cells were significantly elevated in untreated patients, and decreased with piroxicam treatment to subnormal levels. OKT4-staining helper cells were depressed in untreated patients and remained depressed as a result of piroxicam administration (26). In these results, piroxicam appeared to be far more effective than any of the other NSAIA that patients were taking prior to the experiments. Removal of glass-adherent mononuclear cells from peripheral blood samples taken from patients with rheumatoid arthritis did not improve their depressed PHA response *in vitro* (26). This finding indicates that the PG induced suppression of cellular immunity in arthritic patients differs from that described by the same research team for patients with Hodgkin's disease or sarcoidosis (36,37). It is possible, therefore, that the PG-producing dendritic cell described by Dayer et al. (14) or the dendritic cell MCF production, participates in the local activation of suppressor cells in the synovium, rather than the macrophage, as is clearly the case in Hodgkin's disease and sarcoidosis.

AUTOIMMUNE DISEASE

The general term *autoimmunity* applies to conditions where self tolerance breaks down, and an immunological response such as the formation of autoantibody is generated. Autoantibodies are not always disease related, and are occasionally found in normal individuals. The frequency of autoantibody generation increases, for example, with increasing age, or as a result of widespread tissue damage, suggesting a breakdown in self tolerance (38). The term *autoimmune disease* applies in cases where self-directed immune responses have pathological results. Evidence of a breakdown in self tolerance can be seen in a wide variety of autoimmune disorders ranging from organ-specific diseases, such as thyroiditis, to the generation of autoantibody against a wide variety of cell and tissue antigens, as occurs in systemic lupus erythematosis (SLE).

The exact natural causes of such autoimmune diseases are generally unclear; however, many such diseases can be induced by viral infection or by the administration of certain drugs. For example, persistent viral infections have been noted in one animal model of immune complex disease, which occurs in F_1 hybrids of New Zealand black (NZB)×New Zealand white (NZW) mice (39). Such animals develop circulating immune complexes containing antibodies to nuclear antigens and viral antigens. Das has suggested that this represents hyperreactivity of B cells caused by the loss of normal regulatory control by T cells, and may account for the increased synthesis of autoantibody in many autoimmune diseases (39). PGs are probably involved in this defect. A deficiency of PGE_1 and/or TXA_2 synthesis, and an increase in PGE_2 synthesis is known to result in activation of B cells and suppression of T cells, along with enhancement of fibrous-tissue formation (40). Viruses are known to block the enzyme delta-6-desaturase, which is involved in PGE_1 synthesis, thereby depressing cell-mediated immunity (39).

Drugs suspected of contributing to the development of autoimmune disease, such as lithium salts, procainamide, and hydralazine, also appear to block PGE_1 synthesis and/or TXA_2 synthesis resulting in autoantibody formation (41). Colchicine, on the other hand, is known to enhance TXA_2 synthesis and to enhance the biological actions of PGE_1, and this drug has been shown to help in the management of progressive systemic sclerosis, Sjögren's syndrome, Behçet's syndrome, vasculitis, amyloidosis, and scleroderma (41,42).

Steinhauer et al. reported that the synthesis of both PGE_2 and TXB_2 in the liver and kidney tissue of NZB/W F_1 hybrid mice increased with increasing age (43). This increased synthesis coincided with the appearance and progression of the SLE-like disorder characteristic of these mice. Dietary supplementation with histadine and/or zinc suppressed the increase in prostanoid generation, an important observation as it had been reported earlier that administration of histadine or zinc to SLE-prone NZB/W F_1 mice prevented the progression of the disease (43). A defect in T-helper cell activity has also been indicated in some autoimmune diseases such as SLE. In the case of SLE-prone NZB/W F_1 mice, this helper-cell defect appears to be IL-2 related, and can be corrected *in vitro* by the addition of phorbol myristate acetate (44). Furthermore, limiting dilution analysis has shown that the number of IL-2-secreting T-helper cell precursors are not reduced in NZB/W F_1 mice, and yet IL-2 production is severely limited (45). As is discussed in detail in Chapter 4, IL-2 production and the response of lymphocytes to IL-2 are both limited by the presence of PGE, suggesting another route by which this metabolite may contribute to the pathophysiology of SLE.

Immunoregulatory disturbances related to early, massive T-lymphocyte hyperplasia are characteristic of MRL/1pr mice. Such animals usually die from a progressive immune-complex-mediated glomerulonephritis. Administration of 15-methyl PGE_1 to these mice prevents lymphoproliferation and the subsequent renal disease (46). This effect is as a result of the prevention of the age related loss of autologous (Lyl(+)2,3(−)-dependent) MLR and concanavalin-A-induced (Ly1(−)2, 3(+)-dependent) suppressor-cell activity, presumably through maintaining a normal T-lymphocyte subset balance (46). The resistance of autoimmune MRL/1pr mice to *Listeria* infection can also be maintained through the administration of PGE_1 (47).

The effects of PGE_1 administration were studied by Takamori and Ide in rats immunized with acetylcholine receptors to induce experimental autoimmune myasthenia gravis (48). This disease is thought to be mediated by antibody-dependent, complement-mediated cytolysis, and a delayed-type cutaneous hypersensitivity response to acetylcholine receptors. Daily injections of PGE_1 prevented the development of acute myasthenia gravis in this model. Such treatment, however, did not prevent subsequent onset of chronic disease, which is caused by accelerated degradation of acetylcholine receptors by antibody and complement-mediated cell lysis in the postsynaptic membrane (48).

Human patients with autoimmune thyroid diseases were studied by Pacini et al. (49). These investigators developed an assay based on the observations that the addition of indomethacin to PHA-stimulated peripheral blood lymphocytes, increased lymphocyte blastogenesis in normal subjects and, to a smaller degree in diseases with reduced prostaglandin-induced suppressor cell activity. Pacini et al. found that lymphocytes from patients with Graves' disease were essentially unaffected by the addition of indomethacin to the assay *in vitro*. Patients receiving anti-thyroid drugs or radioiodine had a normal response to the addition of indomethacin to their lymphocyte cultures, and the response of lymphocytes from patients with Hashimoto's thyroiditis was quite variable upon the addition of indomethacin. These results suggest that PG-induced suppressor cell activity is a part of the pathology of at least some of these disease processes (49).

Finally, PGE_2 may contribute to the etiology of multiple sclerosis (MS). MS is a disease that can be arrested by the immunosuppressive agent cyclophosphamide, suggesting that multiple sclerosis is an immunologically based disease (50). During periods of acute disease activity, the peripheral blood of MS patients contains activated monocytes, which produce substantial amounts of PGE_2 (51). It is a concern of researchers in this area to determine the role of the PGE_2 produced in modulating the phenotype and suppressor activity of CD8(+) lymphocytes. The observation that CD8 is present on sheep oligodendrocytes has suggested the possibility that an autoimmune response directed against CD8 might directly effect brain cells as well as lymphocytes (52). This possibility was supported by the finding that MS serum contains a lymphocytotoxic factor (53) and the observation that antibody to CD8 caused the disappearance *in vitro* of both CD8 and suppressor activity (54). Attempts to demonstrate CD8 on human oligodendrocytes, however, have thus far been unsuccessful.

LITERATURE CITED

1. Nakamura RM: *Immunopathology: Clinical Laboratory Concepts and Methods.* Little, Brown & Company, Boston, pp 418–58, 1974.
2. Nakamura RM, Tucker ES: Immune-complex diseases. *In*: Ritzmann SE, Daniels JC (eds.), *Serum Protein Abnormalities: Diagnostic and Clinical Aspects.* Alan R. Liss, Inc., New York, pp. 295–330, 1982.
3. Broder I, Urowitz MB, Gordon DA: Appraisal of rheumatoid arthritis as an immune complex disease. *Med. Clin. North Am.* 56:529–39, 1972.
4. Bartfeld H: Distribution of rheumatoid factor activity in nonrheumatoid states. *Ann. N.Y. Acad. Sci.* 168:30–40, 1969.
5. Bravo JF, Herman JH, Smyth CJ: Musculoskeletal disorders after renal homotransplantation. *Ann. Intern. Med.* 66:87–104, 1967.
6. Allen JC, Kunkel HG: Antibody versus gamma-G after repeated blood transfusions in man. *J. Clin. Invest.* 45:29–39, 1966.
7. Abruzzo JL, Christian CL: The induction of rheumatoid factor-like substance in rabbits. *J. Exp. Med.* 114:791–806, 1961.
8. Chayen J, Bitensky L: Lysosomal enzymes in inflammation. *Ann. Rheum. Dis.* 30:522–36, 1971.
9. Harris ED, DiBona DR, Krane SM: Collagenases in human synovial fluid. *J. Clin. Invest.* 48:2104–13, 1969.
10. Harris ED, Evanston JM, DiBona DR, Krane SM: Collagenase and rheumatoid arthritis. *Arthritis Rheum.* 13:83–94, 1970.
11. Robinson DR, McGuire MB, Levine L: Prostaglandins in the rheumatic diseases. *Ann. N.Y. Acad. Sci.* 256:318–29, 1975.

12. Krane SM: Aspects of the cell biology of the rheumatoid synovial lesion. *Ann. Rheum. Dis.* 40:433–48, 1981.
13. Wolinsky SI, Goodwin JS, Messner RP, Williams RC: Role of prostaglandin in the depressed cell-mediated immune response in rheumatoid arthritis. *Clin. Immunol. Immunopathol.* 17:31–7, 1980.
14. Dayer JM, Krane SM, Russell RGG, Robinson DR: Production of collagenase and prostaglandins by isolated adherent rheumatoid synovial cells. *Proc. Natl. Acad. Sci. USA* 73:945–9, 1976.
15. Dayer JM, Passwell JH, Schneeberger EE, Krane SM: Interactions among rheumatoid synovial cells and monocyte-macrophages: production of collagenase-stimulating factor by human monocytes exposed to concanavalin A or immunoglobulin Fc fragments. *J. Immunol.* 124::1712–20, 1980.
16. Krakaner KA, Zurier RB: Pinocytosis in human synovial cells in vitro. Evidence for enhanced activity in rheumatoid arthritis. *J. Clin. Invest.* 66:592–8, 1980.
17. Woolley DE, Brinckerhoff CE, Mainardi CL: Collagenase production by rheumatoid synovial cells: morphological and immunohistochemical studies of the dendritic cell. *Ann. Rheum. Dis.* 38:262–70, 1979.
18. Dayer JM, Russell RGG, Krane SM: Collagenase production by rheumatoid synovial cells: stimulation by a human lymphocyte factor. *Science* 195:181–3, 1977.
19. Dayer JM, Robinson DR, Krane SM: Prostaglandin production by rheumatoid synovial cells. Stimulation by a factor from human mononuclear cells. *J. Exp. Med.* 145:1399–1404, 1977.
20. Dayer JM, Golding SR, Robinson DR, Krane SM: Effects of human mononuclear cell factor on cultured rheumatoid synovial cells. Interactions of prostaglandin E_2 and cyclic adenosine 3′,5′-monophosphate. *Biochim. Biophys. Acta* 586:87–105, 1979.
21. Mochan E, Uhl J, Newton R: Interleukin-1 stimulation of synovial cell plasminogen activator production. *J. Rheumatol.* 13:15–19, 1986.
22. Mochan E, Uhl J, Newton R: Evidence that interleukin-1 induction of synovial cell plasminogen activator is mediated via prostaglandin E_2 and cAMP. *Arthritis Rheum.* 29:107–84, 1986.
23. Koun JH, Halushka PV, LeRoy EC: Mononuclear cell modulation of connective tissue function. Suppression of fibroblast growth by stimulation of endogenous prostaglandin production. *J. Clin. Invest.* 65:543–54, 1980.
24. Dayer JM, Roelke M, Krane SM: Some factors modulating collagenase production by cultured adherent rheumatoid synovial cells. *Clin. Res.* 29:557A, 1981.
25. Ceuppens JL, Rodriguez MA, Goodwin JS: Nonsteroidal antiinflammatory agents inhibit the synthesis of IgM-rheumatoid factor in vitro. *Lancet* i:528–30, 1981.
26. Goodwin JS, Ceuppens JL, Rodriguez MA: Administration of nonsteroidal antiinflammatory agents in patients with rheumatoid arthritis. *J. Am. Med. Ass.* 250:2485–8, 1983.
27. Vaughan JH, Chihara T, Moore TL, Robbins DL, Tanimoto K, Johnson JS,

McMillan R: Rheumatoid factor producing cells detected by direct hemolytic plaque assay. *J. Clin. Invest.* 933–41, 1976.

28. Robinson DR, Levine L: Prostaglandin concentrations in synovial fluid in rheumatic diseases: action of indomethacin and aspirin. *In*: Robinson HJ, Vane JR (eds), *Prostaglandin Synthetase Inhibitors*. Raven Press, New York, pp 223–8, 1975.

29. Tan P, Shore AA, Leary P, Keystone EL: Interleukin abnormalities in recently active rheumatoid arthritis. *J. Rheumatol.* 11:593–6, 1984.

30. McKenna RM, Ofosu-Appiah W, Warrington RJ, Wilkins JA: Interleukin 2 production and responsiveness in active and inactive rheumatoid arthritis. *J. Rheumatol.* 13:28–32, 1986.

31. Waxman J, Lockshin M, Schnapps J, Doneson I. Cellular immunity in rheumatic diseases. *Arthritis Rheum.* 4:499–506, 1973.

32. Lockshin M, Eisenhauer A, Kohn R, Weksler M, Black S, Mushlin S: Cell mediated immunity in rheumatic diseases. *Arthritis Rheum.* 3:245–50, 1975.

33. Silverman H, Johnson J, Vaughan J, McGlamory J: Altered lymphocyte reactivity in rheumatoid arthritis. *Arthritis Rheum.* 3:509–15, 1976.

34. Veys EM, Hermanns P, Schindler J, Kung PL, Goldstein G, Synens J, Van Wanve J: Evaluation of T cell subsets with monoclonal antibodies in patients with rheumatoid arthritis. *J. Rheumatol.* 9:25–9, 1982.

35. Ceuppens JL, Goodwin JS, Searles RP: The presence of Ia antigen on human peripheral blood T cells and T-cell subsets: Analysis with monoclonal antibodies and the fluorescence-activated cell sorter. *Cell. Immunol.* 64:277–92, 1981.

36. Goodwin JS, Messner RP, Bankhurst AD, Peake GT, Saiki J, Williams RC: Prostaglandin producing suppressor cells in Hodgkin's disease. *New Engl. J. Med.* 297:963–8, 1977.

37. Goodwin JS, Dettoratius R, Israel H, Peake GT, Messner RO: Suppressor cell function in sarcoidosis. *Am. Int. Med.* 90:169–79, 1979.

38. Gilliland BG: Introduction to clinical immunology. *In*: Petersdorf RG, Adams RD, Braunwald E, Isselbacker KJ, Martin JB, Wilson JD (eds.), *Harrison's Principles of Internal Medicine*, 10th edition. McGraw Hill, New York, pp. 344–61, 1980.

39. Das UN: Auto-immunity and prostaglandins. *Int. J. Tissue React.* 3:890–94, 1981.

40. Cunnane SC, Manku MS, Horrobin DF: The pineal and regulation of fibrosis. *Med. Hypotheses* 5:403–14, 1979.

41. Horrobin DF: The regulation of prostaglandin biosynthesis. Negative feedback mechanisms and the selective control of formation of 1 and 2 series prostaglandins: relevance to inflammation and immunity. *Med. Hypotheses* 6:687–709, 1980.

42. Das UN: Prostaglandins and immune response in cancer. *Int. J. Tissue React.* 2:4, 1980.

43. Steinhauer HB, Batsford S, Schollmeyer P, Kluthe R: Studies on thromboxane B_2 and prostaglandin E_2 production in the course of murine autoimmune

disease: inhibition by oral histidine and zinc supplementation. *Clin. Nephrol.* 24:63–8, 1985.
44. Santoro TJ, Luger TA, Ravache ES, Smolen JS, Oppenheim JJ, Steinberg AD: In vitro correction of the interleukin 2 defect of autoimmune mice. *Eur. J. Immunol.* 13:601–4, 1983.
45. Hefeneider SH, Conlon PJ, Dower SK, Henney CS, Gillis S: Limiting dilution analysis of interleukin 2 and colony stimulating factor producer cells in normal and autoimmune mice. *J. Immunol.* 132:1863–71, 1982.
46. Eastcott JW, Kelley VE: Preservation of T lymphocyte activity in autoimmune MRL/1pr mice treated with prostaglandin. *Clin. Immunol. Immunopathol.* 29:78–85, 1983.
47. Kelley VE, Wing E: Loss of resistance to *Listeria* infection in autoimmune MRL/1pr mice: protection by prostaglandin E$_1$. *Clin. Immunol. Immunopathol.* 23:705–10, 1982.
48. Takamori M, Ide Y: Effects of prostaglandin E$_1$ in experimental autoimmune myasthenia gravis. *Neurology* 32:410–13, 1982.
49. Pacini F, Fragh P, Mariotti S, DeGroot LJ: Effect of indomethacin on phytohemagglutinin-stimulated peripheral blood lymphocytes in thyroid autoimmune diseases. *J. Clin. Immunol.* 2:335–42, 1982.
50. Hauser SL, Dawson DM, Lehrich JR, Beal MF, Kevy SV, Propper RD, Mills JA, Weiner HL: Intensive immunosuppression in progressive multiple sclerosis. *New Engl. J. Med.* 308:173–80, 1983.
51. Dore-Duffy P, Zurier RB: Lymphocyte adherence in multiple sclerosis. Role of monocytes and increased sensitivity of MS lymphocytes to prostaglandin E. *Clin. Immunol. Immunopathol.* 19:303–13. 1981.
52. Oger JJ-F, Szuchet S, Antel J, Arnason BGW: A monoclonal antibody against human T suppressor lymphocytes binds specifically to the surface of cultured oligodendrocytes. *Nature* (London) 295:66–8, 1982.
53. van den Noorts ST, Jernholm RL: Lymphotoxic activity in multiple sclerosis serum. *Neurology* 21:783–93, 1971.
54. Antel JE, Oger JJ, Jackevicius F, Kuolt S , Arnason BGW: Modulation of T lymphocyte differentiation antigens: potential relevance for multiple sclerosis. *Proc. Natl. Acad. Sci. USA* 79:330–4, 1982.

9

Traumatic injury and surgery

There has been increasing concern for the immunological depression, which is unavoidably acquired as a result of major accidental or operative injury in an otherwise normal host. Immunological changes following injury are precipitous and leave the host vulnerable to life-threatening sepsis. Fortunately, injury-induced immune depression is completely reversed with patient recovery. The critical period, therefore, occurs during acute care where controlling immunity in favor of the host is the goal of both clinical care and research. Most of the lessons, to date, have been learned from thermal injuries. The immunological changes that occur as a result of burns appear to have a great deal in common with the immune consequences of other types of tissue injury.

IMMUNE DEPRESSION IN BURN PATIENTS

It is clear that major thermal injuries often precipitate a profound, multicentric immunological depression, which is thought to predispose patients to sepsis. Impairment of immune function is almost universal in patients with burns over greater than 40% of the body surface area, and in very young or very old patients with far smaller burns. Briefly, the immunological changes, which occur in these patients, include the following:

1. total loss of skin-test reactivity and recall-antigen responses;
2. the release of endotoxin, tissue-degradation products, hormones, cytokines and lymphokines with immunosuppressive properties into the general circulation;
3. activation of the complement system (both classical and alternate pathways) with the production of complement-split products with immunoregulatory capabilities;
4. reduced monocyte/macrophage function with increased suppressor macrophage function, increased immunosuppressive PGE production, and depressed phagocytosis;

171

5. a transient depression in B cell numbers and immunoglobulin production (both primary and secondary responses are affected);
6. depression of neutrophil functions including chemotaxis, phagocytosis, chemiluminescence, and intracellular killing;
7. depletion of fibronectin and serum opsonic activity;
8. decreased natural killer (NK) cell and lymphokine-activated killer (LAK) cell function;
9. long-term and profound depression of T-lymphocyte response with increased T-suppressor cell activity;
10. a reported reversal of T-lymphocyte helper cell/suppressor-cell ratios.

The period of acute burn care is one of particularly diminished macrophage and lymphocyte function, vigorous suppressor T-cell activity and reduced help, the circulation and activity of characterized and uncharacterized immunosuppressive mediators, the presence of leukopenia/lymphopenia, and a deficiency in IL-2 production and IL-2-dependent T-cell activation. The immunological alterations in burn patients have been reviewed in greater detail elsewhere (1-4). Two clinical manipulations seem critical to reversing burn-induced-immune changes: removal of the dead or injured tissue and restoration of the surface barrier to wound colonization through wound closure. Immunological perturbations will persist in burn patients as long as these two conditions are not met. With closure of the burn wound often comes rapid restoration of immunological competence and, frequently, full recovery of the patient. The earlier these clinical manipulations can be accomplished, the earlier immunological restoration will be achieved, thereby decreasing septic threat and increasing patient chances for survival. In addition, experiments aimed at boosting immune responsiveness in burned animals also show a reduction of sepsis, which suggests that similar clinical manipulation of human burns, while not yet attempted, might be useful (5).

TRAUMA-ASSOCIATED IMMUNODEFICIENCY

It is widely held that burns represent a model by which the immunological changes, which occur following other types of traumatic injury and following major surgery (a type of controlled trauma), can be better understood. Indeed, many of the changes that occur in burn patients also occur following blunt or penetrating trauma (6); however, relatively few attempts have been made to study the immunologic responses in patients with such injuries. While research is hampered by the difficulty in grouping patients for study, there is clearly a correlation between a high injury severity score (ISS) and the development of immune depression

(7,8). The exact threshold of injury, which results in loss of immune competence, however, has not been established as it has for burn patients.

As with the burn patient, the most important clinical evidence of immune depression in the trauma patient is the dramatic increase of septic complications, which accompanies increasing injury severity. For example, Baker et al. studied 437 accidental deaths in San Francisco during 1977 and it was found that 78% of late deaths were caused by sepsis (9). Polk has noted a similar incidence for septic death (75%) in patients with major thermal injuries (10). In Baker's study most of the trauma patients who died from sepsis had multiple injuries, and it was hypothesized that many of the deaths could have been prevented by extensive early debridement. As noted above, this procedure has been shown to help restore immunological competence in burn patients (3).

Patients with major traumatic injuries are frequently anergic. In a study of 176 trauma patients (129 with blunt injuries, 27 with thermal injuries, and 20 with penetrating injuries), Christou found that such a lack of skin-test reactivity (anergy), detected upon patient admission, was indicative of frequent sepsis and later patient death (11). A decrease in neutrophil function has been reported to accompany anergy following blunt trauma. In a study of 31 patients suffering from blunt trauma, it was found that 6 patients had normal skin-test reactions with no sepsis or death, while 25 demonstrated skin-test anergy with 20% sepsis and 16% mortality rates (12). Abnormalities of PMN chemotaxis and adherence were detectable within 2 h of injury, and appeared to be related to both anergy and the occurrence of circulating immunosuppressive factors in the blood. No major abnormalities in serum immunoglobulin levels or in complement levels were detected (12). Keane et al. showed that lymphocytes of trauma patients displayed a reduced response to mitogens and alloantigens proportional to the extent of injury, infectious sequelae, and prognosis (13). Keane et al. also reported an increase in suppressor cell activity following multiple trauma in three of the group of four patients studied (7).

In 1984, O'Mahoney et al. compared the T-cell mitogen response, percentage of putative suppressor [OKT8(+)] and helper [OKT4(+)] lymphocytes, circulating suppressor cell activity, and serum suppression of lymphocyte activation in 31 multiple trauma patients with 10 normal controls (8). They found significant suppression in lymphocyte response to mitogens 1–5 days after injury in 12 of the patients, and a shift in the normal OKT4(+)/OKT8(+) T-cell ratio, followed by the appearance of serum suppressive activity in 6 of the 12 patients. Circulating suppressor cell activity (measured by functional assays) was seen early after injury in 3 of 12 patients. O'Mahoney et al. concluded that impaired T-lymphocyte function commonly follows severe multiple injury, and that this was a major factor predisposing patients to sepsis (8).

IMMUNOLOGICAL CHANGES FOLLOWING SURGERY

In light of the immunological changes that occur following injury it is not surprising that patients undergoing major surgical procedures are also prone to develop immunological depression. The first careful study demonstrating this fact was carried out by Slade et al. (14) who measured immune function *in vivo* and *in vitro* following nephrectomy in 12 normal renal transplant donors. All were normal individuals with normal immune responsiveness prior to surgery. Total lymphocyte, B-cell, and T-cell numbers, mitogen and MLR decreased upon induction of anesthesia and continued to fall during (and after) operation. The delayed hypersensitivity response *in vivo* to cutaneous antigen challenge declined more gradually and was still falling at the fifth postoperative day. While these findings clearly indicated immunological compromise in these patients, Slade et al. reported no clinically significant problems with sepsis (14).

Contrary to the conclusions of Slade et al., however, there is evidence that immune depression in surgical patients can have important clinical consequences. Using a standard skin-testing procedure, Christou et al., in a prospective study of 503 skin-test-positive surgical patients, found that 6.4% developed a long-lasting anergy in the postoperative period (15). This anergy group had a 41% rate of sepsis and a 22% mortality rate, compared to 5% sepsis and 3% mortality in the reactive group. Christou concluded that only major surgical procedures (esophagogastrectomy, colectomy, aortic resection, etc.) resulted in anergy (15); however, the extent of surgery required to produce immunological depression and anergy is still in question. Fabricius et al. recently studied patients who underwent simple surgery for inguinal hernia (16), and found a profound suppression of T-lymphocyte function as measured by T-cell colony formation, while conventional tests such as PHA stimulation showed no significant change.

Greco et al. reported immune depression in 85% of the normal elective-surgery patients they studied via a leukocyte migration inhibition assay (17). Abnormalites persisted for 60 days or longer in half of the patients studied. Christou and Meakins reported a larger series of patients in whom skin-test anergy and neutrophil chemotactic response were measured (18), and they concluded that the worst combination for a surgical patient was cutaneous anergy coupled with decreased neutrophil chemotaxis. When they occurred together, these two tests were predictors of a high rate of sepsis and frequent mortality. Christou and Meakins also showed that the serum of anergic patients contained circulating mediators, which were responsible for the depressed neutrophil response. Subsequent attempts to characterize the serum-borne inhibitor of neutrophil chemotaxis showed that it was of low molecular weight (19).

Lymphocyte response is also altered by surgery, an effect primarily confined to T-cell responses (measured *in vitro* via mitogen or antigen stimulation) and mediated by suppressor cell activation (20,21). In addition, the induction of a primary, cellular immune response to dinitrochlorobenzene (DNCB) appeared to be profoundly impaired in cancer patients as a result of surgery (22), also indicating lymphocyte response abnormality.

THE ROLE OF CIRCULATING MEDIATORS

The question of how immunologic depression in the injured patient is mediated has received a great deal of attention. Experimental studies have shown that the general circulation of patients after injury or a major operation, often contains factors affecting vascular permeability, causing hemolytic changes in red blood cells, depressing cardiac output and renal function, causing profound immunosuppression, and, in some cases, precipitating multiple organ failure. Especially interesting have been cross-perfusion experiments in which it has been found that the transfer of blood from one animal after thermal injury, can produce these effects in another, normal animal (23). Ninnemann has shown that the response of normal lymphocytes placed in burn patient serum is often profoundly impaired (Figure 9.1) (24). On the other hand, the response of burn patient lymphocytes removed from the "burn environment" and placed in normal serum, returns toward normal following a period of incubation (25). Warden et al. have made similar observations using neutrophil chemotaxis as the test system (26). Potential causes of surgical immune depression are probably multiple, cumulative, and not easily differentiated. It is becoming clear, however, that arachidonic acid metabolites have a major role in the changes which take place.

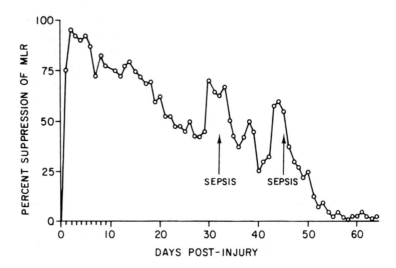

Figure 9.1. Typical suppressive activity of sequential serum samples drawn from patients with major thermal injuries versus lymphocyte response in mixed lymphocyte cultures (MLC). (Reprinted with permission from Ninnemann JL: Immunologic complications associated with thermal injuries. *In*: RI Walker, DF Gruber, TJ MacVittie, and JJ Conklin (eds). *Pathophysiology of Combined Injury and Trauma.* Armed Forces Radiobiology Research Institute, Bethesda, MD, pp. 206–19, 1985.)

INJURY, INFLAMMATION, AND THE PROSTAGLANDINS

The pathophysiology of all forms of injury, including burns, is characterized by a strong inflammatory response, particularly where the wound depth is only partial skin thickness (27,28). Injury is followed by immediate changes in blood flow with the development of tissue edema, an increase in microvascular permeability, an increase in the extravascular osmotic activity of damaged tissue, and finally, an infiltration of the tissues by leukocytes (29,30). There is evidence that histamine mediates the early part of this response (31), while in rats and mice, serotonin (5-hydroxy-tryptamine) may also be important (32). The later phases of burn wound inflammation are mediated by polypeptides of the kinin system (33) and various permeability factors (34), as well as the products of arachidonic acid metabolism.

Experiments by Anggard et al. showed that a scald burn of the dog paw was followed by an efflux of PG (primarily PGE_2) in the lymphatic fluid draining the paw. This suggested increased PGE_2 biosynthesis as a result of the injury (35). Guinea pig studies verified this effect of thermal damage to the skin, and demonstrated an increased excretion of urinary 5-beta,7 alpha-dihydroxy-11-ketotetranor-prostanoic acid, a major metabolite of PGE (36). Lipid extracts of guinea pig skin excised 2 h after injury contained PG-like material at 20–40 times the concentration found in the lipid extracts of skin excised immediately. This observation indicates local, rather than a systemic increase in PG biosynthesis as a result of injury. Results by Hamberg and Jonsson also demonstrated an increase in the biosynthesis of PGF and its metabolites (36).

Arturson et al. reported that smooth muscle activity was induced by acidic lipid extracts of blister fluid from burn patients, and chromatographic procedures demonstrated that this activity was induced by PGE and PGF (37). In a pooled sample of burn blister fluid, the concentration of PGE_2 was found to be 1.9 ng/mL, well above the levels reported to have vasodilatory effects. TXB_2 has also been identified in burn blister fluid at a concentration of 1.7 ng/mL (38). Acidic lipid extracts obtained from burn wound secretions collected between 1 and 3 weeks postinjury in 12 patients were found to have biologic activity equivalent to 5–30 ng/mL of PGE_2 as shown in Figure 9.2 (39). The highest values were observed during the first 24 h. Burn wound extracts were found to contain primarily PGE_1 during the early hours postburn, but later, PGE_2 and $PGF_{2\alpha}$ predominated (40).

Heggers et al. described the direct release of large quantities of PGs from skin cells as a result of thermal injuries (41). Ninnemann and Stockland found that the concentration of PGE in sera from patients with greater than 40% burns was quite high (10–30 ng/mL) and that these same sera were often significantly suppressive to lymphocyte responsiveness *in vitro* (Table 9.1) (42). Ninnemann and Stockland found that both delipidation and the addition of rabbit anti-PGE_2 significantly reduced the lymphocyte suppressive activity of low molecular weight fractions of burn serum (42). These data are shown in Table 9.2. Radioisotope labeling experiments suggested stable, PGE-containing molecular complexes of 68 000 daltons and 5000 daltons, respectively (Figure 9.3) (42). These investigators hypothesized that a carrier might be responsible for the persistence of PGE in the circulation beyond its usual clearance in the lungs (43). The participation of carrier molecules in the physiologic effects of the PGs is not a new concept, and it is generally believed that the transport of PG across biologic membranes (such as is required for the generation of suppressor cells) is carrier mediated, since PG cannot freely diffuse through membrane structures (44).

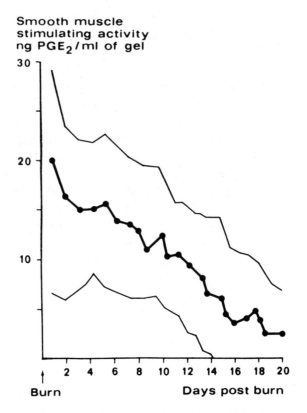

Figure 9.2. The biological activity (equivalent to PGE_2) in nanograms per milliliter of burn wound fluid in 12 patients on different days after burn injury. Mean values and range are shown. (Reprinted with permission from Arturson MG: Discussion: How are prostaglandins and leukotrienes involved in immunological alterations? *J. Trauma* 9(suppl): S131, 1984.)

Grbic et al. found that lymphocytes from burn patients were particularly sensitive to the suppressive effects of PGE_2 (Figure 9.4) (45), an effect similar to that reported by Goodwin et al. who showed that physical stress of many types increased the sensitivity of lymphocytes to suppression by PGE_2 (46) and Rodrick et al. who reported that burn serum was capable of suppressing IL-2 synthesis (47). Wood et al. presented evidence that the suppression of IL-2 production in an animal model of thermal injury was related to PG synthesis (48). Wood et al. also proposed that reduced IL-2 production was a (the?) fundamental immune deficiency in burn patients (49). IL-1 production, on the other hand, was

Table 9.1. *Burn patient serum samples: PGE and cortisol content, and suppressive activity in mixed lymphocyte cultures.*

Patient	Day Post-injury	PGE Content	Cortisol Content	% Sup-pression
Control	—	190 pg/ml	88 ng/ml	—
RD (80% BSA)	4	1,650 pg/ml	230 ng/ml	33
	6	1,400 pg/ml	160 ng/ml	21
	12	1,275 pg/ml	170 ng/ml	74
	30	1,255 pg/ml	240 ng/ml	97
	46	2,000 pg/ml	200 ng/ml	73
	53	1,850 pg/ml	220 ng/ml	90
KH (70% BSA)	3	1,400 pg/ml	86 ng/ml	98
	75	2,700 pg/ml	32 ng/ml	0
WP (47% BSA)	1	1,850 pg/ml	470 ng/ml	100
	11	1,700 pg/ml	190 ng/ml	19
	12	2,350 pg/ml	220 ng/ml	46
	19	1,900 pg/ml	180 ng/ml	69
VE (41% BSA)	1	650 pg/ml	200 ng/ml	10
	2	725 pg/ml	320 ng/ml	95
	3	650 pg/ml	370 ng/ml	80
	4	1,800 pg/ml	360 ng/ml	95

Note: BSA, body surface area.

Source: Reprinted with permission from Ninnemann JL, Stockland AE: Participation of prostaglandin E in immunosuppression following thermal injury. *J. Trauma* 24:201–7, 1984.

found to be at normal or supranormal levels. Wood hypothesized that reduced IL-2 levels reflected the reduction in (IL-2 producing) OKT4(+) helper cells, but significantly, found that a direct correlation was lacking.

Other thermal-injury-related human experiments in this area have been done by Teodorczyk-Injeyan et al. (50,51). This group assessed T-cell proliferation and the expression of Tac (IL-2) receptors postburn in patients with 5–68% total body surface area (TBSA) burns. T-cell dependent Ig synthesis, MLC, and IL-2 production levels were also assessed. There was a marked difference in the study group, in terms of Tac receptor expression, Il-2 production, and lymphocyte response levels, between patients who survived their injuries and those who died (50). Teodorczyk-Injeyan et al. showed that impaired expression of IL-2 receptors in the burn patient was mediated by PGE_2 and could be duplicated in

Table 9.2. *Effect of delipidation on addition of anti-PGE$_2$ on the suppressive activity of prostaglandin-containing burn serum fractions.*

mw 5,000 Fraction of Serum	Day Post injury	Treatment	DPM ± SD	% sp/st
Control	—	None	82,050 ± 2,434	—
	—	Delipidated	68,400 ± 1,763	—
	—	Anti-PGE$_2$	110,001 ± 2,003	—
RD (80% BSA)	12	None	55,650 ± 7,952	32 sp
	12	Delipidated	57,154 ± 12,351	16 sp
	12	Anti-PGE$_2$	111,861 ± 1,425	1 st
RD (80% BSA)	30	None	64,550 ± 6,635	21 sp
	30	Delipidated	58,968 ± 2,426	14 sp
	30	Anti-PGE$_2$	113,942 ± 12,075	3 st
VE (41% BSA)	3	None	57,152 ± 2,234	30 sp
	3	Delipidated	62,641 ± 1,866	8 sp
	3	Anti-PGE$_2$	111,251 ± 1,512	1 st

Note: Significant suppression was observed by additions of RD serum fractions obtained 12 and 30 days postburn and VE serum fractions 3 days postburn ($p<0.01$), but not after delipidation of RD 30 day and VE 3 day fractions, or after the addition of anti-PGE$_2$ to each of the cultures. % sp/st, percent suppression or stimulation; DPM ± SD, disintegrations per min ± standard deviation; BSA, body surface area.

Source: Reprinted with permission from Ninnemann JL, Stockland AE: Participation of prostaglandin E in immunosuppression following thermal injury. *J. Trauma* 24:201–7, 1984

culture via the addition of burn serum (51). Ninnemann also reported that lymphocyte exposure to PGE$_2$ (or burn serum) also results in their inability to respond to exogenous IL-2 stimulation (52).

LEUKOTRIENE PRODUCTION IN BURNS

Braquet et al. investigated the possible involvement of LTs in inflammation and the anergy often seen in burn patients (53). Their results suggested that postburn anergy was related to leukocyte secretion of immunoreactive agents, among them eicosanoids such as LTB$_4$. Furthermore, LTB$_4$ appeared to be a reliable marker of clinical outcome. These studies, however, are preliminary, and it is clear that more work in this area is justified.

THERAPEUTIC REVERSAL OF EICOSANOID SUPPRESSION

Heggers has shown that early cooling of burn injuries in an animal model can prevent conversion of partial-thickness injuries to full-thickness inju-

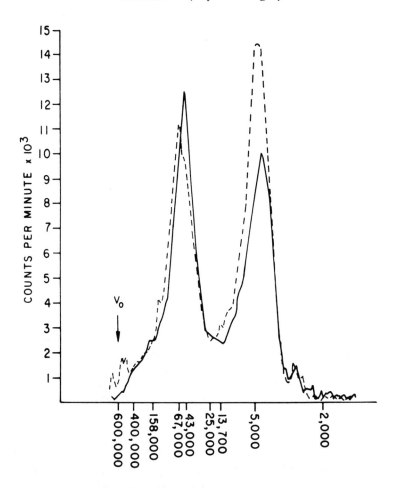

Figure 9.3. The elution, from a Sephadex G-200 column, of [³H]PGE₂ mixed with 0.5 normal human serum (solid line), or with a 0.5 mL sample of serum 1 day after burn from a burn injured patient (broken line). Peaks represent counts per minute of 100 uL aliquots of 1.0 mL fractions. (Reprinted with permission from Ninnemann JL, Stockland AE: Participation of prostaglandin E in immunosuppression following thermal injury. *J. Trauma* 24:201–7, 1984.)

ries (54). This effect is caused by the temperature-dependent suppression of cyclooxygenase activity within the damaged tissue. It appears that injury-associated deterioration of cell-mediated immunity can also be prevented by blocking or reducing prostanoid production. Hansbrough et al. demonstrated that the cellular immune response could be restored in burn-injured animals by means of treatment with various phar-

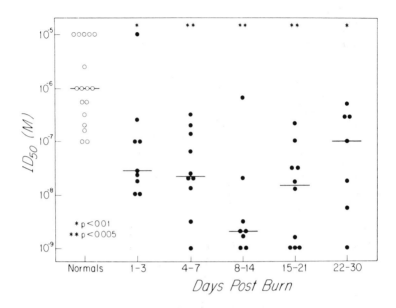

Figure 9.4. Increased sensitivity of lymphocytes from burn patients to inhibition by PGE_2. Each value represents the concentration of PGE_2 that caused 50% inhibition of PHA-induced blastogenesis. The horizontal lines in each group indicate the median value. (Reprinted with permission from Grbic JT, Wood JJ, Jordan A, Rodrick ML, Mannick JA: Lymphocytes from burn patients are more sensitive to suppression by prostaglandin E_2. *Surg. Forum* 108–10, 1985.)

macological agents including the cyclooxygenase inhibitors ibuprofen and indomethacin (55). Finally, Faist et al. demonstrated encouraging improvement of PHA responses in the lymphocytes of trauma patients as a result of indomethacin treatment (56). These data are shown in Figure 9.5.

TRANSFUSION INDUCED IMMUNOSUPPRESSION

It is suspected that many aspects of surgical care contribute to systemic immune depression. There have been many reports dealing with the immunosuppressive effects of various antibiotics and topical agents, multiple anesthetic procedures, and the administration of blood products. The suppressive effects of blood transfusions have been studied using animal models of prolonged allograft survival (57,58) and more recently, a popliteal lymph node assay for graft versus host (GVH) responsiveness

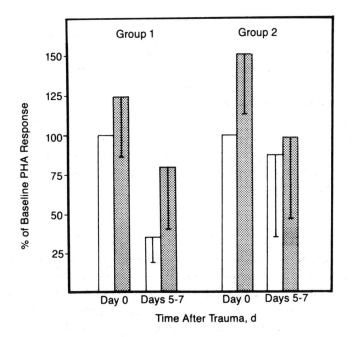

Figure 9.5. Effects of indomethacin treatment on PHA responses in infected patients (group 1) and uninfected patients (group 2) at baseline (day 0) and 5–7 days after injury. Open bars indicate unfractionated cells alone; filled bars represent unfractionated cells treated with indomethacin. Responses are shown as mean percentages (with standard deviations) of baseline responses for untreated cells. (Reprinted with permission from Faist E, Kuper TS, Baker CC, Chaudry IH, Dwyer J, Baue AE: Depression of cellular immunity after major injury. *Arch. Surg.* 121:1000–5, 1986.)

(59). Results of these studies suggest that transfusion results in the induction of suppressor T lymphocytes (57), the production of antiidiotypic antibodies (60), sensitization resulting in clonal deletion (61), and the increased production of PGs such as PGE_2 (62).

Evidence for the importance of PGE_2 production following transfusion has come from the study of dialysis patients who received multiple transfusions in preparation for kidney transplants (63–65). PGE levels have been observed to be quite elevated in these patients. Likewise in mice, it has been reported that the injection *in vivo* of sheep red blood cells resulted in significantly increased immunosuppressive PGE production by splenic macrophages (66) and that the transfusion of allogeneic murine red cells is also immunosuppressive (67). That PGE production was stimulated by this transfer of murine erythrocytes was demonstrated by Shelby et al. (68). They found that postoperative allogeneic (third party) blood transfusion resulted in prolonged survival of MHC-matched heart allografts, and that pretransplant-antigen-specific allogeneic transfusion decreased the GVH response in the mice. Indomethacin blocked this transfusion-induced suppression in both pre- and posttransplant models. Suppression was also blocked by the administration of anti-PGE antibody (68). The mechanism for this transfusion-induced suppression of the immune response appears to be the generation of prostaglandin-induced suppressor cells, through increased PGE_2 production by the macrophage/monocyte (69).

BACTERIAL ENDOTOXIN AND SEPSIS

Circulatory shock, which is related to bacterial sepsis, is a major problem in the surgical patient populations just described. The exact cause remains unclear. It is thought, however, that endotoxins from Gram-negative bacteria often play a major role, and an endotoxin shock model in animals is frequently utilized to study the problem. The participation of a wide variety of humoral mediators in the shock process has been investigated using this model, including kinins, angiotensin, histamine, serotonin, catecholamines, and endogenous opiates (70). In recent years, the eicosanoids have been added to the list and are now recognized as major participants in the pathophysiology of sepsis and of circulatory shock (71–73).

The intravenous injection of endotoxin has been found to result in a very large but transient increase in the plasma concentrations of TXB_2 and 6-keto-$PGF_{1\alpha}$ in a number of animal species (Table 9.3). A small, delayed and more prolonged increase in TXB_2 and 6-keto-$PGF_{1\alpha}$ has been reported in animals with septic shock caused by fecal peritonitis or cecal ligation (74,75). The injection of thromboxane inhibitors blocks a number of physiological changes associated with septic shock, including acute cardiopulmonary changes, increases in serum lysosomal enzymes, the production of fibrin/fibrinogen degradation products, which are themselves immunosuppressive (76), and thrombocytopenia (77). These results suggest that TXA_2 is responsible for many of the pathophysiological effects of endotoxin during shock.

It has also been found that NSAIAs and dietary deficiency of essential fatty acids increase the survival of rats subjected to endotoxin shock (78,79), as well as improving survival in animals with septic shock (80).

Table 9.3. A summary of endotoxin-induced increases in TXB_2 and 6-keto-$PGF_{1\alpha}$ concentrations in various animal models.

Species	TxB₂ Peak (ng/ml)	Time (min)	6-keto-PGF₁ₐ Peak (ng/ml)	Time (min)
Sheep (anesthetised)	8.5[a]	35	0.5[a]	75
Sheep (conscious)	3.24[a]	15[a]	5.0[a]	45[a]
Baboon (conscious)	13.6[a]	15	3.3[a]	120
Baboon (anesthetised)	3.4[a]	5	4.8[a,b]	120
Pony	4.1[a]	15	4.0[a]	120
Goat	17[a]	30	1.4[a]	90
Goat	13.1	30	3.1	90
Pig	1.7	20	(no change)	(no change)
Dog	4.46	5	0.55	5
Cat	0.83	2	2.32	2
Rabbit	–	–	1.63	120
Rat	2.97	30	–	–

[a] Value estimated from graph or calculated from a table
[b] Concentrations still increasing
– profile of metabolite not determined

Source: Reprinted with permission from Ball HA, Cook JA, Wise WC, Halushka PV: Role of thromboxane, prostaglandins, and leukotrienes in endotoxic and septic shock. Intensive Care Med. 12:116–26, 1986.

These results support the idea that other cyclooxygenase products are involved in septic shock, and preliminary studies indicate that lipoxygenase inhibitors have beneficial effects as well (81). Clinical studies have demonstrated increases in TXB_2 and 6-keto-$PGF_{1\alpha}$ concentrations in the plasma of patients with septic shock, and increased LTD_4 in the pulmonary edema fluid of patients with adult respiratory distress syndrome (82,83). This parallels the observations of Ball et al. in animals (Table 9.3).

The effects of bacterial endotoxin on the immune system likewise appear to be mediated by the products of arachidonic acid metabolism.

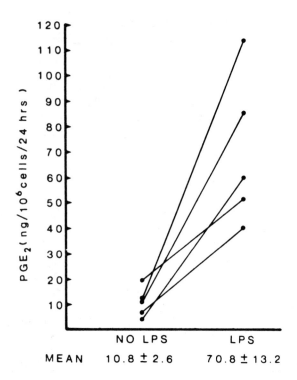

Figure 9.6. Production of PGE_2 by monocytes cultured with and without 20 ng/mL endotoxin. Values represent concentrations of PGE_2 in culture supernatants as measured by radioimmunoassay. Reprinted with permission from Ellner JJ, Spagnuolo PJ: Suppression of antigen and mitogen induced human T lymphocyte DNA synthesis by bacterial lipopolysaccharide: mediation by monocyte activation and production of prostaglandins. *J. Immunol.* 123:2689–95, 1979.)

The work of Ellner and Spagnuolo suggested that macrophages, activated by bacterial lipopolysaccharide, produced immunoreactive PGE_2 (Figure 9.6), which in turn was responsible for endotoxin-induced suppression of T-lymphocyte function (84). The drugs indomethacin and RO-205720 blocked this inhibition of lymphocyte function by endotoxin-activated macrophages. Similar results have been reported by Rietschel (85), Kahn and Bracket (86), and others (87,88).

Studies by Ninnemann et al. indicated that endotoxin is capable of generating nonspecific suppressor cells, that the ability of these cells to suppress is unaffected by mitomycin treatment and that the generation of these cells is PG dependent (89). On the basis of depletion experiments, it

Table 9.4. *The effect of PG synthetase inhibitors upon suppression produced by normal lymphocytes preincubated in endotoxin or in the serum of burn patients.*[a]

Preincubation, ng/ml	Culture		
	Serum alone	Serum + indomethacin	Serum + meclofenamate
E. coli 055:B5			
500	30[b]	88	119
50	43[b]	101	86
5.0	55[b]	85	78
E. coli 0111:B4			
500	53[b]	98	72
50	49[b]	86	95
5.0	52[b]	110	101
P. aeruginosa			
500	28[b]	58	69
50	61[b]	99	93
5.0	61[b]	109	87
S. marcescens			
500	33[b]	93	90
50	51[b]	115	89
5.0	44[b]	145	126
S. minnesota			
500	19[b]	75	65
50	54[b]	93	99
5.0	60[b]	72	125
Patient 1	58[b]	115	85
Patient 2	57[b]	97	94
Patient 3	62[b]	95	98
Patient 4	30[b]	93	82

[a] Data represent the relationship of lymphocyte stimulation to the control level of response and are expressed as the percentage of the control. The 7-day human mixed lymphocyte cultures contained 2×10^5 responder and 2×10^5 mitomycin-treated allogeneic stimulator lymphocytes.
[b] Significantly different from control, $P < 0.01$.

Source: Reprinted with permission from Ninnemann JL, Stockland AE, Condie JT: Induction of prostaglandin synthesis-dependent suppressor cells with endotoxin: occurrence in patients with thermal injuries. *J. Clin. Immunol.* 3:142–50, 1983.

appeared that the macrophage was the source of the PG in this sequence. Co-culture experiments showed that the suppressor cell generated in this sequence was short lived, that is, the suppressive activity of endo-toxin/PG-activated suppressor cells disappeared after 24 h incubation. It was unclear, however, whether it was the suppressor cell which lost activity, or whether there was a loss in the ability of the initiating monocyte/macrophage to produce PG (89). When radioimmunoassay was used to measure PGE_2 in culture supernatants, however, it appeared that the monocyte/macrophages lost their ability to secrete PGs following 24 h in culture (89). Finally, the clinical relevance of these findings was demonstrated by studies with serum and plasma samples from patients with major thermal injuries. *Limulus*-positive patient sera were found to induce similar PG-dependent, short-lived suppressor cell activity (25,89). Like the endotoxin-induced supressor cell activity, this suppression was abrogated by monocyte depletion or by the addition of PG synthetase inhibitors (Table 9.4). With some burn sera, however, suppression was not affected by either monocyte depletion or the addition of cyclooxygenase inhibitors. This Ninnemann et al. interpreted as being a result of the presence of PGs (particularly PGE_2) in these same burn plasma. Radioimmunoassay revealed high concentrations of PGE_2, ranging from 600 to 2700 pg/mL, in these plasma samples (89).

LITERATURE CITED

1. Ninnemann JL (ed.), *The Immune Consequences of Thermal Injury.* Williams & Wilkins, Baltimore, 1981.
2. Ninnemann JL: Immunologic defenses against infection: alterations following thermal injuries. *J. Burn Care Rehabil.* 3:355–66, 1982.
3. Ninnemann JL (ed.), *Traumatic Injury: Infection and Other Immunologic Sequelae.* University Park Press, Baltimore, 1983.
4. Rapaport FT, Milgrom F, Kano K, Gesner B, Solowey HC, Casson PR, Silverman HI, Converse JM: Immunologic sequelae of thermal injury. *Ann. N.Y. Acad. Sci.* 15:1004–8, 1968.
5. Hansbrough JF, Zapata-Sirvent R, Petersen V, Bender EM, Claman HN: Modulation of suppressor cell activity and improved resistance to infection in the burned mouse. *J. Burn Care Rehabil.* 6:270–4, 1985.
6. Howard RJ: Effect of burn injury, mechanical trauma, and operation on immune defenses. *Surg. Clin. No. Am.* 59:199–211, 1979.
7. Keane RM, Munster AM, Birmingham W: Suppressor cell activity after major injury. *J. Trauma* 22:770–3, 1982.
8. O'Mahoney JB, Palder SB, Wood JJ, McIrvine A, Rodrick ML, Demling RH, Mannick JA: Depression of cellular immunity after multiple trauma in the absence of sepsis. *J. Trauma* 24:869–75, 1984.
9. Baker CC, Oppenheimer L, Stephens B, Lewis FR, Trunkey DD: Epidemiology of trauma deaths. *Am. J. Surgery* 140:144–50, 1980.
10. Polk HC: Consensus summary on infection. *Trauma* 19 (suppl): 894–6, 1979.

11. Christou NV: The delayed hypersensitivity skin test response, granulocyte function, and sepsis in surgical patients. *J. Burn Care Rehabil.* 6:157–66, 1985.

12. Christou NV, Meakins JL: Phagocytic and bactericidal functions of poly-morphonuclear neutrophils from anergic surgical patients. *Can. J. Surg.* 25:444–8, 1982.

13. Keane RM, Birmingham W, Shatney CM, Winchurch RA, Munster AM: Prediction of sepsis in the multi-traumatic patient by assays of lymphocyte dysfunction. *Surg. Gynecol. Obstet.* 156:163–7, 1983.

14. Slade MS, Simmons RL, Yunis E, Greenburg LJ: Immunodepression after major surgery in normal patients. *Surgery* 78:363–72, 1975.

15. Christou NV, Superina R, Broadhead M, Meakins JL: Postoperative depression of host resistance: determinants and effect of peripheral protein-sparing therapy. *Surgery* 92:786–92, 1982.

16. Fabricius E, Stahn R, Fabricius HA: Funktionsstorungen des thy-musabhangingen Immune systems nach operativen Engriffen. *Fortschr. Med.* 98:7630–3, 1980.

17. Greco RS, Dick L, Duckenfeld J: Perioperative suppression of the leukocyte migration inhibition assay in patients undergoing elective operations. *Surg. Gyn. Obstet.* 147:717–20, 1978.

18. Christou NV, Meakins JL: Neutrophil function in surgical patients: two inhibitors of granulocyte chemotaxis associated with sepsis. *J. Surg. Res.* 26:355-64, 1979.

19. Christou NV, Meakins JL: Partial analysis and purification of poly-morphonuclear neutrophil chemotactic inhibitors in serum from anergic patients. *Arch. Surg.* 118:156–60, 1983.

20. Salo M: Effect of anesthesia and open heart surgery on lymphocyte responses to phytohemagglutinin and concanavalin A. *Acta Anesth. Scand.* 22:471–9, 1978.

21. Wang BS, Heacock EH, Wu AVO, Mannick JA: Generation of suppressor cells in mice after surgical trauma. *J. Clin. Invest.* 66:200–9, 1980.

22. Tarpley JL, Twomey PL, Catalona WJ, Chretien PB: Suppression of cellular immunity by anesthesia and operation. *J. Surg. Res.* 22:195–201, 1977.

23. Baxter CR, Moncrief JA, Prager MD: A circulating myocardial depressant factor in burn shock. *In:* Matter P, Barclay TL, Konickova Z (eds.), *Research in Burns.* Haus-Huber Verlag, Bern, pp. 499–509, 1971.

24. Ninnemann JL: Suppression of lymphocyte response following thermal injury. *In:* Ninnemann JL (ed.), *The Immune Consequences of Thermal Injury.* Williams & Wilkins, Baltimore, pp. 66–89, 1981.

25. Ninnemann JL, Stein MD, Condie JL: Lymphocyte response following thermal injury: the effect of circulating immunosuppressive substances. *J. Burn Care Rehabil.* 2:196-9, 1981.

26. Warden GD, Mason AD, Pruitt BA: Suppression of leukocyte chemotaxis in vitro by chemotherapeutic agents used in the management of thermal injuries. *Ann. Surg.* 181:363–9, 1975.

27. Spector WG, Willoughby DA: The inflammatory response. *Bacteriol. Rev.* 27:117–54, 1963.

28. Zweifach BW, Grant L, McCluskey RT (eds.): *The Inflammatory Process*. Academic Press, New York, 1965.
29. Arturson MG: Arachidonic acid metabolism and prostaglandin activity following burn injury. *In*: JL Ninnemann (ed.), *Traumatic Injury: Infection and Other Immunologic Sequelae*. University Park Press, Baltimore, pp. 57–78, 1983.
30. Spector WG, Willoughby DA: Experimental suppression of the acute inflammatory changes of thermal injury. *J. Path. Bact.* 78:121–32, 1959.
31. Eldery H, Lewis GP: Kinin-forming activity and histamine in lymph after tissue injury. *J. Physiol.* 169:568–83, 1963.
32. Sparrow EM, Wilhelm DL: Species differences in susceptibility to capillary permeability factors: histamine, 5-hydroxytryptamine and compound 48/80. *J. Physiol.* 137:51–65, 1957.
33. Rocha e Silva M, Antonio A: Release of bradykinin and the mechanism of production of a thermic edema in the rat's paw. *Med. Exp.* 3:371–82, 1960.
34. Tasaki I: Mechanisms in the delayed and prolonged vascular permeability changes in inflammation. *Kumamoto Med. J.* 21:1–12,, 1968.
35. Anggard E, Arturson G, Jonsson CE: Efflux of prostaglandins in lymph from scalded tissues. *Acta Physiol. Scad.* 80:46A–7A, 1970.
36. Hamberg M, Jonsson CE: Increased synthesis of prostaglandins in guinea pig following scalding injury. *Acta Physiol. Scand.* 87:240–5, 1973.
37. Arturson G, Hamberg M, Jonsson CE: Prostaglandins in human burn blister fluid. *Acta Physiol. Scand.* 87:270–6, 1973.
38. Jonsson CE, Granstrom E, Hamberg M: Prostaglandins and thromboxanes in burn injury in man. *Scand. J. Plast. Reconstr. Surg.* 13:45–7, 1979.
39. Arturson G: Prostaglandins in human burn wound secretion. *Burns* 3:12–8, 1977.
40. Jonsson CE, Shimizu Y, Friedholm BB, Granström E, Oliw E: Efflux of cyclic AMP, prostaglandin E_2 and F_2 and thromboxane B_2 in leg lymph of rabbits after scalding injury. *Acta Physiol. Scand.* 107:377–84, 1979.
41. Heggers JP, Loy GL, Robson MC, DelBeccaro EJ: Histological demonstration of prostaglandins and thromboxanes in burned tissue. *J. Surg. Res.* 28:110–17, 1980.
42. Ninnemann JL, Stockland AE: Participation of prostaglandin E in immunosuppression following thermal injury. *J. Trauma* 24:201–7, 1984.
43. Lee JB: Are the prostaglandins hormones? *In*: JB Lee (ed.), *Perspectives on the Prostaglandins*. Medcom Press, New York, p. 18, 1973.
44. Bito LZ, Wallenstein M, Baroody R: The role of transport processes in the distribution and disposition of prostaglandins. *Adv. Prostgl. Thrombox. Res.* 1:297–303, 1976.
45. Grbic JT, Wood JJ, Jordan A, Rodrick ML, Mannick JA: Lymphocytes from burn patients are more sensitive to suppression by prostaglandin E_2. *Surg. Forum* 108–10, 1985.
46. Goodwin JS, Bramberg S, Staszak C, Kaszubowski PA, Messner RP, Neal JF: Effect of physical stress on sensitivity of lymphocytes to inhibition by prostaglandin E_2. *J. Immunol.* 127:518–22, 1981.
47. Rodrick ML, Saporoschetz IB, Wood JJ, Davis CF, Mannick JA: Serum

suppression of interleukin-2 (IL-2) production and IL-2 action in thermal injury. *Surg. Forum* 98–9, 1985.

48. Wood JJ, Rodrick ML, Grbic JT, Mannick JA: Suppression of interleukin-2 (IL-2) production in an animal model of thermal injury is related to prostaglandin synthesis. *Proc. Surg. Infect. Soc.* p. 33, 1986.

49. Wood JJ, Rodrick ML, O'Mahoney JB, Palder SB, Saporoschetz I, D'eon P, Mannick JA: Inadequate interleukin 2 production: a fundamental immunological deficiency in patients with major burns. *Ann. Surg.* 200:311–20, 1984.

50. Teodorczyk-Injeyan J, Sparkes BG, Mills GB, Peters WJ, Falk RE: Impairment of T cell activation in burn patients: a possible mechanism of thermal injury-induced immunosuppression. *Clin. Exp. Immunol.* 65:570–81, 1986.

51. Teodorczyk-Injeyan J, Sparkes BG, Peters WJ, Gerry K, Falk RE: Prostaglandin E-related impaired expression of interleukin-2 receptor in the burn patient. *Adv. Prost. Thrombox. Leukotr. Res.* 17:147–50, 1987.

52. Ninnemann JL: PGE regulates lymphocyte stimulation by IL-2. *Proc. Am. Burn Assn.* 19:10, 1987.

53. Braquet M, Dicousso R, Garay R, Guilband J, Carsin H, Braquet P: Leukotriene secretion in burn-injured patients: a relation with anergy? *In:* JW Streilein, F Ahmad, S Black, B Blomberg, RW Voellmy (eds.), *Advances in Gene Technology: Molecular Biology of the Immune System.* Cambridge University Press, New York, pp. 117–8, 1986.

54. Heggers JP, Robson MC: Prostaglandins and thromboxanes. *In:* JL Ninnemann (ed.), *Traumatic Injury: Infection and Other Immunologic Sequelae.* University Park Press, Baltimore, pp. 79–102, 1983.

55. Hansbrough J, Peterson V, Zapata-Sirvent R, Claman HN: Postburn immunosuppression in an animal model: II. Restoration of cell-mediated immunity by immunomodulating drugs. *Surgery* 95:290–5, 1984.

56. Faist E, Kupper TS, Baker CC, Chaudry IH, Dwyer J, Baue AE: Depression of cellular immunity after major injury. *Arch. Surg.* 121:1000–5, 1986.

57. Shelby J, Wakely E, Corry RJ: Suppressor cell induction in donor specific transferred mouse heart recipients. *Surgery* 96:296–300, 1984.

58. Shelby J, Fick J, Kolegraff RJ, Wakely E, Corry RJ: Prolonged heart allograft survival induced by massive preoperative blood transfusions. *Transplantation* 40:113-4, 1985.

59. Michel F, Rouquette AM, Ansgner JC, Ponsot Y, Thimbault PH: Effect of immunosuppressive drugs upon the transfusion effect. *Transplant. Proc.* 17:2425–7, 1985.

60. Burlingham WJ, Sparks-Mackety EMF, Wendel T, Pan MH, Sondel PM, Sollinger HW: Beneficial effect of pretransplant donor-specific transfusions: evidence for an idiotype network mechanism. *Transplant. Proc.* 17:2376–9, 1985.

61. Terasaki PI: The beneficial transfusion effect on kidney graft survival attributed to clonal deletion. *Transplantation* 37:119–25, 1984.

62. Keown DA, Descamps B: Improved renal allograft survival after blood transfusion: a nonspecific, erythrocyte-mediated immunoregulatory process? *Lancet* i:20–2, 1979.

63. Jackson V, Tsakus D, Tonner E, Briggs JD, Junor BJR: In vitro prostaglandin E production following multiple blood transfusions in dialysis patients. *Transplant Proc.* 17:2386–89, 1985.
64. Roy R, Lachance JG, Beaudoin R, Grose JH, Noel R: Prostaglandin-dependent suppressor factor induced following 1–5 blood transfusions: role in kidney graft outcome. *Transplant. Proc.* 17:2382–5, 1985.
65. Lenhard V, Gensa D, Opelz G: Transfusion-induced release of prostaglandin E_2 and its role in the activation of T suppressor cells. *Transplant. Proc.* 17:2380–2, 1985.
66. Webb DR, Osheroff PL: Antigen stimulation of prostaglandin synthesis and control of immune responses. *Proc. Natl. Acad. Sci. USA* 73: 1300–4, 1976.
67. Heslop BF, Heslop HE: Allogeneic red blood cells fail to induce haemagglutinating antibodies or cellular alloimmunity in rats and are immunosuppressive. *Transplantation* 28:144-8, 1979.
68. Shelby J, Marushack MM, Nelson EW: Prostaglandin production and suppressor cell induction in transfusion-induced immune suppression. *J. Trauma*, 1988 (in press).
69. Wymack JP, Gallon L, Barcelli U, Alexander JW: Effect of blood transfusions on macrophage arachidonic acid metabolism. *Proc. Surg. Infect. Soc.* p. 41, 1986.
70. Parratt JR: Neurohumoral agents and their release in shock. *In:* Altura BM, Lefer AM, Shumer W (eds.), *Handbook of Shock and Trauma.* Raven Press, New York, pp. 311–36, 1983.
71. Bult H, Herman AG: Prostaglandins and circulatory shock. *In:* Herman AG, Vanhoutte PM, Denolin H, Goossens A (eds.), *Cardiovascular Pharmacology of Prostaglandins.* Raven Press, New York, pp. 327–45, 1982.
72. Fletcher JR: The role of prostaglandins in sepsis. *Scand. J. Infect. Dis.* (Suppl) 31:55–60, 1982.
73. Lefer AM: Role of prostaglandins and thromboxanes in shock states. *In:* Altura BM, Lefer AM, Schumer W (eds.), *Handbook of Shock and Trauma.* Raven Press, New York, pp. 355–76, 1983.
74. Funk MP, Gardiner WM, Roethal R, Fletcher JR: Plasma levels of 6-keto-$PGF_{1\alpha}$ but not TXB_2 increase in rats with peritonitis due to cecal ligation. *Circ. Shock* 16:297–305, 1985.
75. Butler RR, Wise WC, Halushka PV, Cook JA: Gentamicin and indomethacin in the treatment of septic shock: effects on prostacyclin and thromboxane A_2 production. *J. Pharmacol. Exp. Therap.* 225:94–101, 1983.
76. Edgington TS, Curtiss LK, Plow EF: A linkage between the hemostatic and immune systems embodied in the fibrinolytic release of lymphocyte suppressive peptides. *J. Immunol.* 134:471–7, 1985.
77. Ball HA, Cook JA, Wise WC, Halushka PV: Role of thromboxane, prostaglandins and leukotrienes in endotoxic and septic shock. *Intensive Care Med.* 12:116–26, 1986.
78. Ziboh VA, Vanderhoek JY, Lands WEM: Inhibition of sheep vesicular gland oxygenase by unsaturated fatty acids from skin of essential fatty acid deficient rats. *Prostaglandins* 5:233–40, 1974.

79. Cook JA, Wise WC, Butler RR, Reines HD, Rambo W, Hallushka PV: The potential role of thromboxane and prostacyclin in endotoxin and septic shock. *Am. J. Emergency Med.* 2:28–37, 1984.

80. Butler RR, Wise WC, Halushka PV, Cook JA: Thromboxane and prostacyclin production during septic shock. *Adv. Shock Res.* 7:133–45, 1982.

81. Hagmann W, Keppler D: Leukotriene antagonists prevent endotoxin lethality. *Naturwissenschaften* 69:594–5, 1982.

82. Reines HD, Halushka PV, Cook JA, Wise WC, Rambo W: Plasma thromboxane concentrations are raised in patients dying with septic shock. *Lancet* ii:174–5, 1982.

83. Deby-Dupont G, Radoux L, Haas M, Larbuisson R, Noel FX, Lamy M: Release of thromboxane B_2 during adult respiratory distress syndrome and its inhibition by nonsteroidal antiinflammatory substances in man. *Arch. Int. Pharmacodyn.* 259:317–19, 1982.

84. Ellner JJ, Spagnuolo PJ: Suppression of antigen and mitogen induced human T lymphocyte DNA synthesis by bacterial lipopolysaccharide: mediation by monocyte activation and production of prostaglandins. *J. Immunol.* 123:2689–95, 1979.

85. Reitschel ET, Schade U, Luderitz O, Fisher H, Peskar BA: Section I. Prostaglandins in endotoxicosis. *In*: Schlessinger D (ed.), *Microbiology-1980.* American Society for Microbiology. Washington D.C., pp. 66–72, 1980.

86. Kahn A, Brachet E: Involvement of prostaglandins in the local action of endotoxin. *Prostagland. Med.* 6:23–8, 1981.

87. Fletcher JR, Ramwell PW: E. coli endotoxin shock in the baboon: treatment with lidocaine or indomethacin. *In*: Galli C, Galli G, Procellati G (eds.), *Advances in Prostaglandin and Thromboxane Research*, Vol. 3. Raven Press, New York, pp. 183–92, 1978.

88. Flynn JT: Endotoxin shock in the rabbit: the effects of prostaglandin and arachidonic acid administration. *J. Pharmacol. Exp. Therap.* 206:555–66, 1978.

89. Ninnemann JL, Stockland AE, Condie JT: Induction of prostaglandin synthesis-dependent suppressor cells with endotoxin occurrence in patients with thermal injuries. *J. Clin. Immunol.* 3:142–50, 1983.

10

Allergy

Atopic or allergic diseases affect 10–20% of the population of the United States. Genetic factors play an important role in the susceptibility to these diseases, and association with specific HLA haplotypes has been demonstrated. Allergic responses are immunological reactions mediated through the activity of IgE. As early as 1921, Prausnitz and Kustner recognized that serum from allergic individuals contained a humoral sensitizing factor, which they called *reagin* (1). With the advances made in immunochemistry during the 1950s and 1960s came the characterization of reaginic antibody as a fifth distinct immunoglobulin class, IgE (2). While the serum of normal individuals contains only nanogram quantities of IgE, concentrations in allergic individuals can be quite significant.

A variety of cells synthesize membrane glycoproteins that can bind the Fc portion of specific immunoglobulins, and such Fc receptor interactions initiate specific cellular functions (3). Fc receptors on mast cells and basophils have a high affinity for monomeric IgE. Human basophils normally have 10^3–10^6 of these receptors per cell; however, basophils from individuals with high levels of serum IgE have an increased number of receptors (4). As shown in Figure 10.1, the interaction of IgE with surface receptors on basophils and mast cells (as well as some other nonspecific stimuli) results in the release of biologically active mediators such as histamine (5). Histamine is formed in the cells by the cytoplasmic enzyme histidine decarboxylase (6), and although basophils make up less than 1% of the circulating leukocyte pool, they contain all the histamine found in human blood. Basophils are short lived (less than 2 weeks), but mast cells are longer lived. Mast cells are larger than basophils and are capable of releasing more histamine (7).

Histamine is stored in the lysosomes of basophils and mast cells, and released via exocytosis (degranulation) upon stimulation of the cells. Histamine causes its maximal effects within 1–2 min. Inactivation of

194

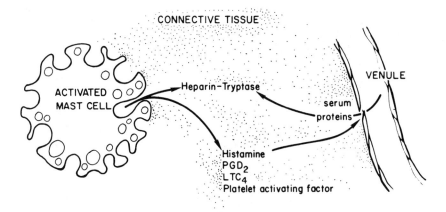

Figure 10.1. Mast-cell-derived vasopermeability factors facilitate the interaction of serum proteins with mast-cell-neutral protease at the interface between connective tissue and the postcapillary venule. (Reprinted with permission from: Schwartz LB: The mast cell. *In*: Kaplan AP (ed.), *Allergy*. Churchill Livingstone, New York, pp. 53–92, 1985.)

histamine occurs rapidly *in vivo* by deamination (through deamine oxidase or histaminase) or methylation (N-methyltransferase). Histamine acts on target organs through H_1 and H_2 receptors. H_1 receptors occur primarily on bronchioles and vascular smooth muscle cells, and H_2 receptors occur primarily on gastric parietal cells. Histamine receptors are also present on lymphocytes (particularly suppressor T cells) and on basophils. It is thought that such histamine receptors mediate a feedback control of the mast cell/basophil secretory system, as well as play a role in regulating the immune response (8).

ARACHIDONIC ACID RELEASE BY BASOPHILS AND MAST CELLS

During the secretion of histamine by basophils and mast cells, a number of important changes occur in the phospholipid metabolism of these cells. There is an increase in the turnover of phosphatidylinositol, phosphatidylcholine, and phosphatidic acid. The activation of basophils and mast cells also results in the release of arachidonic acid from the cellular phospholipids, primarily phosphatidylcholine and phosphatidylethanolamine (9). Some work has suggested that arachidonic acid is released from phosphatidylcholine as a result of the activation of phospholipase

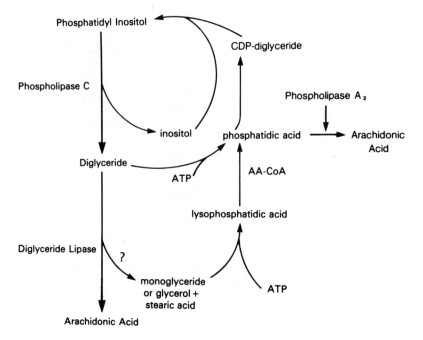

Figure 10.2. Possible pathway for the generation of arachidonic acid from phosphatidylinositol. (Reprinted with permission from Siraganian RP: Biochemical events in basophil/mast cell activation and mediator secretion. *In*: Kaplan AP (ed.), *Allergy*. Churchill Livingstone, New York, pp. 53–92, 1985.)

A_2 (10); however, phosphatidic acid has been suggested as an alternative substrate (11). As proposed in Figure 10.2, 1,2-diacylglycerol may be cleaved by diglyceride lipase to release free arachidonate and 1-mono-acylglycerol. A second product of this cleavage is another 1-mono-acylglycerol or lysophatidylcholine, either of which can fuse membranes and might play a role in the fusion of cell membrane granules (7). Unfortunately, however, the exact metabolic sequence and the identity of the phospholipase enzymes in basophils and mast cells have not yet been determined.

The release of arachidonic acid parallels the release of histamine upon IgE or calcium ionophore stimulation of the cell (7), and requires the presence of Ca^{2+} in the medium. Blocking phospholipase activity in the cells also blocks histamine release, indicating that phospholipase activation is central to cell activation, and not a byproduct of the secretory process (12).

The arachidonic acid released from the mast cell/basophil is metabolized, in part, via the cyclooxygenase pathway, yielding primarily PGD_2 (13), with smaller amounts of $PGF_{2\alpha}$ and TXA_2. The urine of patients with systemic mastocytosis has been found to contain large amounts of 9-alpha-hydroxy-11,15-dioxo-2,3,4,5-tetranorprostane-1,20-dioic acid, a major PGD_2 metabolite, which confirms that PGD_2 is the major cyclooxygenase product in these patients (14). While the addition of PGD_2 to either rat or human mast cells seems to have no effect (15,16), its addition to human basophils enhances the immunological release of histamine (17). Other biological effects of PGD_2 include bronchoconstriction of human and canine airways (18,19), and chemokinesis of neutrophils (20). PGD_2 is also a potent inhibitor of platelet aggregation (21). The systemic release of PGD_2 together with histamine in patients with mastocytoma causes rhinorrhea, bronchorrhea, and hypotension (14). Inhibitors of the cyclooxygenase pathway have been reported (a) to have either little or no effect on histamine release, indicating that the products of cyclooxygenase activity are not involved in the secretory process (10), or (b) to augment histamine release, presumably through diversion of arachidonate to the lipoxygenase pathway (22,23).

SLOW-REACTING SUBSTANCE OF ANAPHYLAXIS AND THE LIPOXYGENASE PATHWAY

The leukotrienes LTC_4, LTD_4, and LTE_4 are major lipoxygenase products of the mast cell/basophil. Demonstration of the biological activity of these compounds antedated their chemical identification by decades. The term *slow-reacting substance of anaphylaxis* (SRS-A) was applied to the uncharacterized active component of the perfusates of sensitized guinea pig lungs following antigen challenge. SRS-A was found to induce a dose-dependent, sustained contraction of guinea pig ileum smooth muscle *in vitro* (24). This same effect could be demonstrated using diffusates of human lung fragments (25). SRS-A was found to be distinct from histamine, since H_1 antihistamines could not inhibit contractile activity (24). In 1977, SRS-A activity was linked to the presence of arachidonic acid metabolites (26). Finally, after the structural identification of the LTs (27), several never-before-observed LTs were found to be responsible for SRS-A activity (28,29). In 1979, SRS-A was found to contain 5-hydroxy-6-S-glutathionyl-7,9,11,14-eicosatetraenoic acid (LTC_4) (28). Subsequently, two additional components of SRS-A were found to be generated from LTC$_4$ by the sequential removal of glutamine to form LTD$_4$ (30) and then glycine to form LTE$_4$ (31).

LTC$_4$, LTD$_4$, and LTE$_4$ cause contraction of airway smooth muscle in systems both *in vitro* and *in vivo* (32). This activity is not suppressed by antihistamines, seems to be preferential for certain types of smooth muscle, and appears to be important in the etiology of bronchial asthma. For example, isolated natural SRS-A and also synthetic sulfidopeptide LTs have a preferential contractile effect on guinea pig lung parenchyma over tracheal strips *in vitro* (33,34). LTC$_4$ and LTE$_4$ are 100 times more active in this effect than histamine, and LTD$_4$ is 1000 times more potent. When SRS-A or synthetic LTs are administered intravenously to guinea pigs, an increase in peripheral airway resistance results that is of much longer duration than that elicited by histamine or PGF$_{2\alpha}$ (32,35). Similar results have also been observed in humans.

HISTAMINE-INDUCED SUPPRESSOR CELLS

Melmon and co-workers showed that histamine-receptor-bearing lymphocytes are capable of modulating both T-cell and B-cell functions (36). Studies of the suppression of lymphocyte proliferation *in vitro* by histamine-activated T cells have shown that monocytes are involved in both the generation and the activity of these cells (37,38). Furthermore, the activity of histamine-activated suppressor cells requires the generation of a lymphokine, histamine-induced suppressor factor (HSF) (38). The activity of this lymphokine can be blocked by the addition of indomethacin to lymphocytes during effector stages, but not during the induction of these specialized cells (37).

In analyzing these data, Rocklin et al. suggested that HSF may act upon a PG-producing cell, presumably a monocyte, to stimulate PG production and thereby suppress lymphocyte response (39). That this is indeed the case is indicated by data generated by Rocklin et al. and summarized in Figure 10.3. Monocytes, but not lymphocytes, incubated with culture supernatants containing HSF increased their production of PGE$_2$, PGF$_{2\alpha}$, and TXB$_2$ by 169, 53, and 49% respectively, and significantly suppressed lymphocyte proliferation (39).

Thus, in summary, it appears that histamine, released from mast cells and basophils, indirectly participates in the regulation of lymphocyte response through the activation of suppressor cells, which regulate PG production by the monocyte, through the elaboration of the lymphokine HSF.

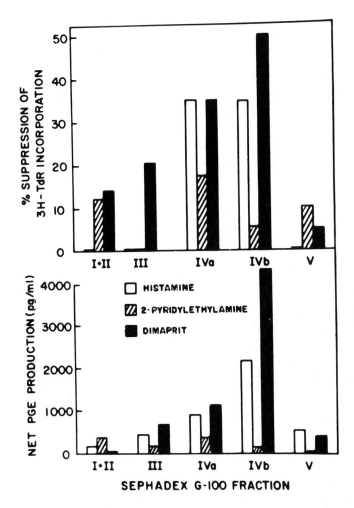

Figure 10.3. Serum-free HSF was generated with histamine, an H_1 agonist (2-pyridylethylamine) or an H_2 agonist (dimaprit) and separated by gel filtration on Sephadex G-100. The fractions (I–V) were assayed in parallel for suppression of [^{3}H]thymidine incorporation, and augmentation of PGE_2 production. Both histamine and dimaprit (but not 2-pyridylethylamine) generated supernatants that inhibited proliferation and augmented PGE_2 production. (Reprinted with permission from Rocklin RE, Kiselis I, Beer DJ, Rossi P, Maggi F, Bellanti JA: Augmentation of prostaglandin and thromboxane production in vitro by monocytes exposed to histamine-induced suppressor factor (HSF). *Cell Immunol.* 77:92–8, 1983.)

Table 10.1. *Arachidonic acid metabolites from biological fluids in humans.*

Disease or Model System	Biological Fluid	AA Metabolite	Comment
Nasal antigen challenges	Nasal secretions	PGD_2, $LTC_4/D_4/E_4$	16 of 17 allergic individuals positive
Nasal challenge with cold, dry air	Nasal secretions	PGD_2, ($LTC_4/D_4/E_4$)	12 patients who report nasal symptoms after cold air exposure; quantitatively less than after antigen
Eye antigen challenge	Tears	$LTC_4/D_4/E_4$	Not separated by HPLC
Skin antigen challenge	Blister fluid	LTC_4	—
Asthma	Sputum	LTC_4, LTB_4	Preliminary data
Asthma	Blood	$PLTC_4/D_1$	Preliminary: 1 of 5 and 2 of 6 patients positive
Cystic fibrosis	Sputum	LTD_4, (LTB_4)	16 of 25 patients positive
Neonatal hypoxemia with pulmonary hypertension	Lung lavage fluid	LTC_4/D_4	5 of 5 patients positive
Adult respiratory distress syndrome	Lung lavage fluid	LTD_4	10 patients; no difference from control in LTC_4, LTB_4, PGD_2, or TXB_2
Psoriasis	Epidermal tissue	AA, 12-HETE, PGE_2, $PGF_{2\alpha}$	AA and 12-HETE much higher in involved than in noninvolved skin
Inflammatory arthritis (rheumatoid arthritis, spondyloarthritis)	Synovial fluid	LTB_4, 5-HETE	24 patients; elevated in rheumatoid arthritis and spondyloarthritis
	Synovial tissue (RA patients only)	5-HETE	7 patients; LTB_4 level normal
Inflammatory bowel disease	Colonic mucosa	LTB_4	12 patients; increased synthesis of AA metabolites from exogenous AA, too.

Source: Reprinted with permission from Peters SP: The cyclooxygenase and lipoxygenase pathways and inflammatory mediators. *In:* Kaplan AP (ed.), *Allergy.* Churchill Livingstone, New York, p. 121, 1985.

CLINICAL PRESENTATION OF ALLERGY

While an IgE response occurs in all normal individuals, the presence of specific immune response genes results in the abnormal IgE-mediated atopic state. The expression of this atopy is also dependent upon antigen specificity, the dose, route, and timing of antigen exposure, drug therapy, and concomitant disease. The atopic (allergic) state is expressed clinically in a variety of ways; however, it is not known why the target organs vary between individuals.

Anaphylaxis is the systemic expression of immediate hypersensitivity. The triggering antigen is usually introduced parenterally, such as by a bee sting or by the injection of penicillin; however, oral introduction of antigen sometimes also results in anaphylaxis. In anaphylaxis there is immediate mast cell activation and degranulation, which results in massive mediator release, and the clinical symptoms of bronchospasm, urticaria, and shock.

Allergic rhinoconjunctivitis is the most common atopic disorder, which also has a pathophysiology resembling immediate hypersensitivity. In this case, IgE is produced locally in the mucosa of the nose and eye after exposure to an airborne antigen, which induces mast cell degranulation and a local anaphylaxis of the nasal and conjunctival membranes.

Urticaria and angioedema are cutaneous forms of immediate hypersensitivity, which are often self limiting and frequently related to the ingestion of specific foods or drugs.

Asthma is an immediate hypersensitivity reaction in the lung. Antigens may be inhaled or ingested, resulting in often isolated preliminary manifestations. In other cases, the precipitating agent is unknown.

Gastrointestinal allergy is an immediate hypersensitivity disease localized in the gastrointestinal tract, with a very poorly understood etiology. Nausea, abdominal cramps, vomiting, and diarrhea may follow within minutes after the ingestion of specific foods.

Atopic dermatitis is a very common eczematous cutaneous eruption, often associated with asthma and allergic rhinitis. It is usually induced by environmental antigens and is frequently associated with very high levels of IgE. As with asthma, atopic dermatitis is associated with a beta-adrenergic-receptor defect.

Clearly these conditions are not all understood to the same degree, however, the underlying pathophysiology involves the sequence of events outlined earlier in this chapter, including the involvement of products of arachidonic acid metabolism, such as PGD_2 and the LTs as summarized in Table 10.1. While mast cell/basophil activation is clearly linked with

Table 10.2. *Some strategies for controlling the effects of arachidonic acid metabolites.*

Substrate (AA) depletion, by dietary or other means

Inhibit release of AA from esterified sources

Inhibition of specific enzymatic reactions (cyclooxygenase, lipoxygenase, leukotriene A synthetase, thromboxane B_2 synthetase, etc.)

Specific receptor or end organ antagonism

Source: Reprinted with permission from Peters SP: The cyclooxygenase and lipoxygenase pathways and inflammatory mediators. *In*: Kaplan AP (ed.), *Allergy*. Churchill Livingstone, New York, p. 122, 1985.

the synthesis and release of these compounds, recent work by Rocklin et al. has shown that mononuclear cells from atopic individuals show a decreased sensitivity to $PGF_{2\alpha}$ and PGD_2 (40). They found that, in contrast to the blastogenic response of normal mononuclear cells, which were inhibited in a dose-dependent manner by both PGE_2 and PGD_2, the cells isolated from atopic individuals were not suppressed by PGE_2, and were inhibited by only high concentrations (10^{-6} M) of PGD_2. Both atopic T-helper and T- suppressor cells exhibited decreased responsiveness to PGE_2 and PGD_2 as compared with normal, nonatopic cells (40). These results suggest that the apparent inability of atopic patient mononuclear cells to regulate immune and inflammatory reactions may be caused by a reduced responsiveness of regulatory subsets to the normal homeostatic influence of PGE_2 and PGD_2.

Although some strategies have been proposed (Table 10.2), unfortunately our knowledge of the PGs and LTs has, until now, had little effect on the clinical management of immediate hypersensitivity. This is because defining the roles of arachidonic acid metabolites is complicated by two important considerations. First, in most physiological situations it appears that a variety of PGs, LTs, and TXs are released at the same time. Some metabolites inhibit, and some metabolites augment, the activity of others. For example, dermal edema is induced by a mixture of LTB_4 and PGE_2, but not by either compound individually (41). Second, in immediate hypersensitivity and inflammation, LTs, for example, are known to act together with nonarachidonate mediators as well, including histamine, the kinins, and complement components. Some of these mediators affect LT synthesis, and share biological properties with them, making it difficult to attribute a response with a specific mediator (42). The development and application of inhibitors of specific metabolites or metabolite groups will greatly aid in the future dissection of pathophysiological effects of individual eicosanoids.

ASTHMA AND ARACHIDONATE METABOLISM

Asthma is a common chronic lung disorder, which shows a familial predisposition, and an association with underprivileged urban areas, colder climates, and industrialized communities (43). Unfortunately the role of specific chemical mediators in this allergic disease are poorly understood. There are suggestions that the histamine released from activated mast cells in the lung is of little importance as the therapeutic administration of antihistaminic drugs is of little benefit (43). It has also been suggested that cyclooxygenase products of arachidonic acid metabolism are also not involved as primary mediators, as the administration of aspirin and aspirinlike drugs is also ineffective (44). In fact, in a small number of patients, it has been found that NSAIAs will actually provoke severe asthmatic attacks (45), supposedly by stimulating the arachidonic acid cascade to favor lipoxygenase products. However, very few studies have been done of local or circulating PG concentrations in asthma patients. Those that are available are usually limited to a single assay utilizing systemic venous blood (46,47), which provides little information as PGs are rapidly metabolized in the lung (in this case, the same organ which produces them). Using such a flawed system, it has been reported that PGE_2 and $PGF_{2\alpha}$ concentrations in asthma patients are not significantly increased from normal (48,49). Chaintreuil et al., however, used a different sampling technique and reported significant increases in PGE_2 and $PGF_{2\alpha}$ (Figure 10.4), which correlated with the degree of bronchial obstruction (50).

Robinson et al. conducted studies *in vitro* and showed that immunologic or calcium-dependent activation of proteolytically dispersed human lung cells, containing 5% mast cells, released large amounts of PGD_2 and TXB_2 (51). The presence of mast cells was required for the release of these metabolites, though it appeared that the primary cellular source was the monocyte/macrophage. Robinson et al. also found that the inhalation of very low concentrations of PGD_2 by subjects with asthma resulted in bronchoconstriction, while much higher concentrations were required to produce the same effect in normal subjects (51). Mathe et al. found that asthmatic patients were approximately 1000 times more sensitive than normal to inhaled $PGF_{2\alpha}$ (52).

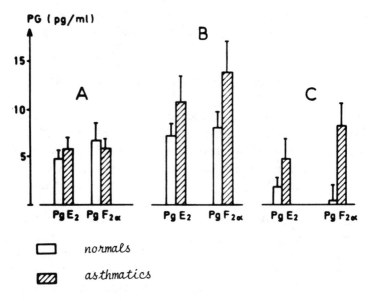

Figure 10.4. PGE_2 and $PGF_{2\alpha}$ concentrations in the plasma of normal subjects (N = 14) and asthmatic patients (N = 21); (A) venous blood, (B) arterial blood, and (C) arteriovenous gradient. (Reprinted with permission from Chaintreuil J, Godard P, Chaintreuil E, Cluzel AM, Crastes de Paulet A, Michel FB: PGE_2 and PGF_2 concentrations in patients with atopic asthma. *In*: Ramwell P (ed.), *Prostaglandin Synthetase Inhibitors: New Clinical Applications*. Alan R. Liss, New York, pp. 45–63, 1980.)

When lung fragments from asthmatic individuals are incubated with antigen *in vitro*, LTC_4, LTD_4 and LTE_4 are also released (53). Pretreatment with a lipoxygenase inhibitor, however, inhibited both LT synthesis and antigen-induced bronchial contraction. Further implication of 5-lipoxygenase products in the pathophysiology of asthma has also been obtained using a Rhesus monkey asthma model. When aerosolized antigen is introduced in sensitized animals, there is an increase in pulmonary resistance, an increase in respiratory rate, and a decrease in dynamic compliance (54). When the NSAIA ETYA was added to the aerosolized antigen, all three effects were greatly reduced. ETYA blocks both the cyclooxygenase and lipoxygenase pathways. ETYA was also able to block the C5a induced smooth muscle contraction of isolated guinea pig trachea (55); however, the cyclooxygenase inhibitor acetylsalicylate, had no effect. FPL-55712, a selective inhibitor of SRS-A, almost completely blocked the C5a response. Such studies suggest the importance of the LTs in the etiology of antigen-induced asthma.

FOOD INTOLERANCE

PGs appear to have cytoprotective effects in the upper bowel, and are released in greater than normal concentrations in patients with peristaltic disorders and diarrhea. PG synthesis inhibitors often prevent symptoms of food intolerance, as well as relieve symptoms in irritable bowel syndrome (56). Likewise, inflammation of the colon is associated with increased PG and TX production, and it is thought that these metabolites contribute to the inflammatory, secretory, and motility dysfunctions in colitis (57).

LITERATURE CITED

1. Prausnitz D, Kustner H: Studien uber die Ueberempfindlichkeit. *Zentralbl. Bakteriol* (A) 86:160–9, 1921.
2. Ishizaka K, Ishizaka T, Hornbrook MM: Physiochemical properties of reaginic antibody V. Correlation of reaginic activity with gamma E-globulin antibody. *J. Immunol.* 97:840–7, 1960.
3. Unkeless JP, Fleit H, Mellman IS: Structural aspects and heterogeneity of immunoglobulin Fc receptors. *Adv. Immunol.* 31:247–70, 1981.
4. Malveaux FJ, Conroy MD, Adkinson NF, Lichtenstein LM: IgE receptors on human basophils relationship to serum IgE concentration. *J. Clin. Invest.* 62:176–86, 1978.
5. Ishizaka T, Ishizaka K: Biology of immunoglobulin E. *Prog. Allergy* 19:60–121, 1975.
6. Padawer J: The mast cell and immediate hypersensitivity. *In*: MK Bach (ed.), *Immediate Hypersensitivity: Modern Concepts and Developments*. Marcel Dekker, New York, pp. 301–67, 1978.
7. Siraganian RP: Biochemical events in basophil/mast cell activation and mediator secretion. *In*: AP Kaplan (ed.), *Allergy*. Churchill Livingstone, New York, pp. 31–51, 1985.
8. Saxon A: Immediate hypersensitivity: approach to diagnosis. *In*: GJ Lawlor, TJ Fischer (eds.), *Manual of Allergy and Immunology: Diagnosis and Therapy*. Little, Brown, Boston, pp. 15–22, 1981.
9. Crews FT, Morita Y, Hirata F, Axelrod J, Siragarian RP: Phospholipid methylation affects immunoglobulin E mediated histamine and arachidonic acid release in rat leukemic basophils. *Biochem. Biophys. Res. Commun.* 93:42–9, 1980.
10. McGivney A, Morita Y, Crews FL, Hirata F, Axelrod JT, Siragarian RP: Phospholipase activation in the IgE-mediated and Ca++ ionophore A23187-

induced release of histamine from rat basophilic leukemia cells. *Arch. Biochem. Biophys.* 212:57-2-80, 1981.

11. Kennerly DA, Sullivan TJ, Sylwester P, Parker LW: Diacylglycerol metabolism in mast cells: a potential role in membrane fusion and arachidonic acid release. *J. Exp. Med.* 150:1039–44, 1979.

12. Siraganian RP: Histamine secretion from mast cells and basophils. *Trends Pharmacol. Sci.* 4:432–7, 1983.

13. Lewis RA, Soter NA, Diamond PT: Prostaglandin D_2 generation after activation of rat and human mast cells with anti-IgE. *J. Immunol.* 129:1627–32, 1982.

14. Roberts LJ, Sweetman BJ, Lewis RA: Increased production of prostaglandin D_2 in patients with systemic mastocytosis. *New Eng. J. Med.* 303:1400–4, 1980.

15. Lewis RA, Holgate ST, Roberts LJ: Preferential generation of prostaglandin D_2 by rat and human mast cells. *In*: Becker EL, Simon AS, Austen KF (eds.), *Biochemistry of the Acute Allergic Reactors.* Alan R. Liss, New York, pp. 239–54, 1981.

16. Lichtenstein LM, Schlemer RP, MacGlashan DW, Peters SP, Schulman ES, Proud D, Creticos PS, Naclerio RM, Kagey-Sobotka A: In vitro and in vivo studies of mediator release from human mast cells. *In*: Kay AB, Austen KF, Lichtenstein LM (eds.), *Asthma: Physiology, Immunopharmacology, and Treatment.* Academic Press, New York, pp. 1–18, 1984.

17. Lichtenstein LM, Gillespie E: The effects of H1 and H2 antihistamines on "allergic" histamine release and its inhibition by histamine. *J. Pharmacol. Exp. Ther.* 192:441–50, 1975.

18. Holgate ST, Church MK, Cushley MJ, Robinson C, Mann JS, Howarth PH: Pharmacologic modulation of airway calibre and mediator release in human models of bronchial asthma. *In*: Kay AB, Austen KL, Lichtenstein LM, (eds.), *Asthma: Physiology, Immunopharmacology, and Treatment.* Academic Press, New York, pp. 391–415, 1984.

19. Wasserman MA, Ducharme DW, Griffin RL: Bronchopulmonary and cardiovascular effects of prostaglandins D_2 in the dog. *Prostaglandins* 13:255–69, 1977.

20. Goetzl EJ, Weller PF, Valone FH: Biochemical and functional basis of the regulatory and protective roles of the human eosinophil. *In*: Weissmann G, Samuelsson B, Paoletti R (eds.), *Advances in Inflammation Research*, Vol. 1. Raven Press, New York, p. 157–67, 1979.

21. Mills DC, MacFarlane DE: Stimulation of human platelet adenylate cyclase by prostaglandin D_2. *Thromb. Res.* 5:401–12, 1974.

22. Marone G, Sobotka AK, Lichtenstein LM: Effects of arachidonic acid and its metabolites on antigen-induced histamine release from human basophils in vitro. *J. Immunol.* 123:1669–77, 1979.

23. Marquardt DL, Nicolotti RA, Kennerly DA, Sullivan TJ: Lipid metabolism during mediator release from mast cells: studies on the role of arachidonic acid metabolism in the control of phospholipid metabolism. *J. Immunol.* 127:845–9, 1981.

24. Brocklehurst WE: The release of histamine and formation of a slow-reacting substance of anaphylaxis (SRS-A) during anaphylactic shock. *J. Physiol.* (London) 151:416–35, 1960.

25. Orange RP, Murphy RC, Karnovsky ML, Austen KF: The physicochemical characteristics and purification of slow reacting substance of anaphylaxis. *J. Immunol.* 110:760–70, 1973.

26. Jakschik B, Falkenheim S, Parker CW: Precursor role of arachidonic acid in slow reacting substance release from rat basophilic leukemic cells. *Proc. Natl. Acad. Sci. USA* 74:4577–81, 1977.

27. Borgeat P, Samuelsson B: Arachidonic acid metabolism in polymorphonuclear leukocytes: Unstable intermediates in formation of dihydroxy acids. *Proc. Natl. Acad. Sci. USA* 76: 3213–7, 1979.

28. Murphy RC, Hammarström S, Samuelsson B: Leukotriene C: a slow-reacting substance from murine mastocytoma cells. *Proc. Natl. Acad. Sci. USA* 76:4275–9, 1979.

29. Morris HR, Taylor GW, Piper PJ, Tippins JR: Structure of slow-reacting substance of anaphylaxis from guinea-pig lung. *Nature* (London) 285:104–6, 1980.

30. Morris HR, Taylor GW, Piper PJ, Tippins JR: The structure elucidation of slow-reacting substance of anaphylaxis (SRS-A) from guinea-pig lung. *Nature* (London) 285:104–6, 1980.

31. Lewis RA, Drazen JM, Austen KF, Clark DA, Correy EJ: Identification of the C(6)-S-conjugate of leukotriene C_4 with cysteine as a naturally occurring slow reacting substance of anaphylaxis (SRS-A): importance of the 11-cis-geometry for biological activity. *Biochem. Biophys. Res. Commun.* 96:271–7, 1980

32. Schwartz LB: The mast cell. *In:* Kaplan AP (ed.), *Allergy.* Churchill Livingstone, New York, pp. 53–92, 1985.

33. Drazen JM, Austen KF, Lewis RA: Comparative airway and vascular activities of leukotrienes C_1 and D in vivo and in vitro. *Proc. Natl. Acad. Sci. USA* 77:4354–8, 1980.

34. Drazen JM, Lewis RA, Wasserman SI: Differential effects of a partially purified preparation of slow-reacting substance of anaphylaxis on guinea pig tracheal spirals and parenchymal strips. *J. Clin. Invest.* 63:1–5, 1979.

35. Drazen JM, Austen KF: Effects of intravenous administration of slow-reacting substance of anaphylaxis, histamine, bradykinin, and prostaglandin F_2 on pulmonary mechanics in the guinea pig. *J. Clin. Invest.* 53:1679–85, 1974.

36. Melmon KL, Rocklin RE, Rosenkranz RP: Autocoids as modulators of the inflammatory and immune response. *Am. J. Med.* 71:100–6, 1981.

37. Beer DJ, Rosenwasser LJ, Dinarello CA, Rocklin RE: Cellular interactions in the generation and expression of histamine-induced suppressor activity. *Cell. Immunol.* 69:101–12, 1982.

38. Rocklin RE, Breard J, Gupta S, Good RA, Melmon KL: Characterization of the human blood lymphocytes that produce a histamine-induced suppressor factor (HSF). *Cell. Immunol.* 51:226–37, 1980.

39. Rocklin RE, Kiselis I, Beer DJ, Rossi P, Maggi F, Bellanti JA: Augmentation of prostaglandin and thromboxane production in vitro by monocytes exposed to histamine-induced suppressor factor (HSF). *Cell. Immunol.* 77:92–8, 1983.

40. Rocklin RE, Thistle L, Audera C: Decreased sensitivity of atopic mononuclear cells to prostaglandin E_2 (PGE_2) and prostaglandin D_2 (PGD_2). *J. Immunol.* 135:2033–9, 1985.

41. Bray MA, Ford-Hutchinson AW, Smith MJH: Leukotriene B_4: an inflammatory mediator in vivo. *Prostaglandins* 22:213–22, 1981.

42. Stetson WF, Parker CW: *Leukotrienes.* Year Book Medical Publishers, Chicago, pp. 175–99, 1984.

43. Reed CE: Asthma. *In*: Kaplan AP (ed.), *Allergy.* Churchill Livingstone, New York, pp. 367–416, 1985.

44. Peters SP: The cyclooxygenase and lipoxygenase pathways and inflammatory mediators. *In*: Kaplan AP (ed.), *Allergy.* Churchill Livingstone, New York, pp. 111–30, 1985.

45. Spector SL, Wangaard CH, Farr RS: Aspirin and concomitant idiosyncrasies in adult asthmatic patients. *J. Allergy Clin. Immunol.* 64:500–6, 1979.

46. Allegra J, Trautlein J, Demers L, Field J, Gillin M: Peripheral plasma determination of prostaglandin E in asthmatics. *J. Allergy Clin. Immunol.* 58:546–50, 1976.

47. Nemoto T, Aoki H, Ike A, Yamada K, Kondo T, Kobayashi S, Inagawa T: Serum prostaglandin levels in asthmatic patients. *J. Allergy Clin. Immunol.* 57:89–94, 1976.

48. Dry J, Pradalier A, Dray F: Dosage des prostaglandines chez les asthmatiques. *N. Presse Med.* 7:4153–4, 1978.

49. Punnonen K, Tammivaara R, Uotila P: Eicosanoid precursor fatty acids in plasma phospholipids and arachidonate metabolism in polymorphonuclear leukocytes in asthma. *In*: Bailey JM (ed.), *Prostaglandins, Leukotrienes, and Lipoxins.* Plenum Press, New York, pp. 555–63, 1985.

50. Chaintreuil J, Godard P, Chaintreuil E, Cluzel AM, Crastes de Paulet A, Michel FB: PGE_2 and PGF_2 concentrations in plasma of patients with atopic asthma. *In*: Ramwell P (ed.), *Prostaglandins Synthetase Inhibitors: New Clinical Applications.* Alan R. Liss, New York, pp. 45–63, 1980.

51. Robinson C, Hardy CC, Holgate ST: Pulmonary synthesis, release, and metabolism of prostaglandins. *J. Allergy Clin. Immunol.* 76:265–71, 1985.

52. Mathe A, Hedqvist P, Holmgren A, Svanborg N: Bronchial hyperreactivity to prostaglandin F_2 and histamine in patients with asthma. *Brit. Med. J.* 1:193–6, 1973.

53. Dahlen SE, Hansson G, Hedquist P, Bjö T, Granström E, Dahlen B: Allergen challenge of lung tissue from asthmatics elicits bronchial contraction that correlates with the release of leukotrienes C_4, D_4, and E_4. *Proc. Natl. Acad. Sci. USA* 80:1712–6, 1980.

54. Patterson R, Harris KE: Inhibition of immunoglobulin E-mediated antigen-induced monkey asthma and skin reactions by 5,8,11,14-eicosatetraynoic acid. *J. Allergy Clin. Immunol.* 67:146–52, 1981.

55. Regal JF, Pickering RJ: Mediation of increased vascular permeability after complement activation. *J. Immunol.* 126:313–6, 1981.
56. Lessof MH, Auderson JA, Youlten LJ: Prostaglandins in the pathogenesis of food intolerance. *Ann. Allergy* 51:249–50, 1983.
57. Zipser RD, Patterson JB, Kao HW, Hauser CJ, Locke R: Hypersensitive prostaglandin and thromboxane response to hormones in rabbit colitis. *Am. J. Physiol.* 249:G457–63, 1985.

INDEX

accessory cells, 34
acetylcholine receptors in myasthenia
    gravis, 166
acute cellular inflammation, 97, 100
acute rejection
    PGE as indicator, 145
    role of platelets, 139
    role of prostacyclin, 140
adult respiratory distress syndrome
    (ARDS), 185
aging, 82, 87, 164
allergic rhinoconjunctivitis, 201
allergy (see also atopy)
    clinical presentation, 201
    general, 194–209
allograft prolongation by pros-
    tacyclin, 139–140
allograft response
    blood flow, 141
    cell mediated immunity, 143
    effect of diet, 148
    granulocyte response, 143
    immune cell infiltration, 137
    mouse skin, 137
    phospholipase activity, 145
    prostacyclin synthetase activity,
        145
    role of calcium, 136
    role of PGs and LTs, 137, 139
allograft survival
    effect of transfusions, 182
    platelet aggregation, 136
amyloidosis, 165
anaphylaxis, 201
anergy
    following thermal injury, 180
    following trauma, 173
    in surgical patients, 174, 175
angioedema, 201

antibody dependent cytotoxicity
    (ADCC)
    in cancer patients, 123, 127
    general, 85
anti-DNA antibody, 157
antigen presentation, 36–38
anti-HLA antibody, maternal, 149
antiidiotype antibody following trans-
    fusion, 183
antirejection, arachidonic acid meta-
    bolites, 136, 138
antitumor immune response, 113
arachidonic acid cascade, 12
arginase, 128
ascites, murine MCDV-12, 121, 123
asthma
    antigen-induced, 205
    etiology, 203
    general, 201–205
    mast cell participation, 203
    pathophysiology, 204
    prostaglandin concentrations, 203
    Rhesus monkey model, 204
    role of C5a, 204, 205
    role of monocyte/macrophage, 203
atopic dermatitis, 201
atopic disease
    association with HLA haplotypes,
        194
    etiology, 194
    incidence, 194
atopy (see also allergy)
    general, 87
    lymphocyte regulatory subsets, 202
    suppressor T cells, 87
autoantibody, 164
autoimmune disease
    definition, 164
    effect of hydralazine, 165

210

stimulation by Ca ionophore, 196
synthesis, 97
vascular effects, 97
histamine-induced suppressor factor
(HSF), 198, 199
histidine, effect on PG production,
165
histidine decarboxylase, 194
HLA antigens on placenta, 149
HLA haplotypes, association with
atopic disease, 194
Hodgkin's disease, 113, 114,
120–121
hydrocortisone
effect on IgM production, 79
effect on IL-2 production, 72
hydroperoxidase, 115
hydroperoxide, 21
hyepracute graft rejection, 139
hypercalcemia, 113

Ia antigen expression, 117
immediate hypersensitivity, 202
immune cell infiltration of grafts,
137
immune complex
disease, 165
general, 45, 157
glomerulonephritis, 166
leprosy, 157
immune depression
rheumatoid arthritis, 158, 161
surgical patients, 174, 175
thermal injuries, 171, 172
threshold of injury, 173
transfusion-induced, 182
trauma, 172–174
immune response, overview, 3–5
immunoglobulin
aggregated IgG, 159
concentrations
following thermal injury, 172
following traumatic injury, 173
Fc fragments, 159
Fc receptor, 194
general, 157
histamine relationship with IgE,
196
production
effect of glucocorticoids, 78, 79
effect of LTs, 44, 46, 82
IgE, 194
immunological interactions, 7–9

immunological recognition, effect of
PGs in graft rejection, 136
immunological surveillance, 114, 118,
121
immunologically privileged sites, 149
inflammation
acute cellular, 97, 100
burn wound, 176, 180
chemical mediators, 97
chronic inflammatory disease, 100,
157, 161
colon, 205
general, 3, 46, 55, 61, 97, 106,
157, 176, 177, 202
proinflammatory metabolites, 100
role of mast cell, 97
role of PGs, 98
synovium, 158
inflammatory bowel disease, 104
interferon
effect on monocyte/macrophage, 36
effect on natural killer (NK) cells,
85, 127
general, 8
IL-2 dependency, 73
LT induction, 73
interleukin 1 (IL-1)
general, 8
MCF similarity, 137
production
effect of LPS, 70, 71
following injury, 178
monocyte/macrophage, 70, 74
interleukin 2 (IL-2)
effect of hydrocortisone, 72
general, 8
LTB_4 as substitute, 73
production
effect of PGE, 166
following thermal injury, 172,
178, 179
rheumatoid arthritis, 161
SLE, 165
receptors following thermal injury, 178
irritable bowel syndrome, 205
ischemia of grafts, 141

Jurkat cells, 53

kallikrein–kinin system, 3
kidney (*see also* renal)
allograft rejection, 148